The Private Security State?

Kirstie Ball, Ana Canhoto, Elizabeth Daniel, Sally Dibb,
Maureen Meadows and Keith Spiller

The Private Security State?

Surveillance, Consumer Data and the War on Terror

Kirstie Ball, Ana Canhoto, Elizabeth Daniel, Sally Dibb, Maureen Meadows and Keith Spiller
The Private Security State? – Surveillance, Consumer Data and the War on Terror

1. edition 2015

Cover: SL grafik (slgrafik.dk)
Typeset: SL grafik
Print: Eurographic Danmark A/S

ISBN: 978-87-630-0332-2

CBS Press
Rosenørns Allé 9
1970 Frederiksberg C
slforlagene@samfundslitteratur.dk
cbspress.dk

Contents

Acknowledgements

There are a number of individuals and organisations without whom this work would not have been possible. First we would like to thank the Leverhulme Trust for funding the work. Thank you for putting your faith in a team of business academics to shed light on the important questions of surveillance, security and the private sector. Second we would like to thank all of the anonymous participants in the fieldwork who so freely gave of their time to share their views with us. We would also like to thank the many industry associations who helped to facilitate access to different participants. In particular we would like to thank Richard Taylor at *Travel Weekly*, Doreen McKenzie of the ABTA Airlines Group, Timon Molloy at *The Money Laundering Bulletin* and the Institute of Money Laundering Prevention Officers who allowed us to attend their events.

We would like to express our gratitude to the many academic friends and colleagues who have engaged with this work throughout the course of its life and have generously given us encouragement as well as their views on its development. In particular we would like to thank David Lyon, The New Transparency Project and Joan Sharpe, without which this work would not have been conceived. We would also like to thank Steve Graham who suggested the title for the book. We would like to thank many of our colleagues who, directly or indirectly, influenced the direction of the work: Anthony Amicelle, Louise Amoore, Caroline Clarke, Sara Degli Esposti, Marieke DeGoede, MariaLaura DiDomenico Gilles Favarel, Kevin Haggerty, Gerry Hanlon, Ben Hardy, Richard Holti, Jef Huysmans, Gavin Jack, David Knights, Reinhard Kreissl, David Murakami Wood, Mike McCahill, Clive Norris, Andras Pap, Jason Pridmore, Michael Nagenborg, Charles Raab, Caroline Ramsey, Bence Ságvarí, John Storey, Pinelopi Troulinnou, David C Wilson, William Webster, Dean Wilson and Nils Zurawski. We would also like to thank the participants in the 4th and 5th Biannual Surveillance and Society Conferences, Meiji University (Dis)embodiment 'unconference' attendees in December 2013, various Living in Surveillance

Societies EU-COST network events, the members of the IRISS and SurPRISE FP7 consortia and the CRISP and Surveillance Studies Centre Doctoral Training workshops. Finally we would like to thank Stewart Clegg and his colleagues at the Copenhagen Business School Press for encouraging us to write this book.

List of abbreviations

ABTA: Association of British Travel Agents
AML/CTF: Anti-Money Laundering/Counter Terror Finance
AML: Anti-Money Laundering
APIS: Advanced Passenger Information Systems
ARA: Asset Recovery Agency (now part of SOCA)
BA: British Airways
BAR UK: Board of Airline Representatives for the United Kingdom
BBA: British Bankers Association
BV: Best Value
CD: Compact Disc
CDD: Customer Due Diligence
CRM: Customer Relationship Management
CSR: Corporate Social Responsibility
DNA: Deoxyribose Nucleic Acid
DVLA: Driver and Vehicle Licensing Agency
EU: European Union
FATF: Financial Action Task Force
FISA: Foreign Intelligence Surveillance Act
FSA: Financial Services Authority
G-7: The Group of Seven Industrialised Nations
GDS: Global Distribution System
HMC: Home Affairs Committee
II: Information Infrastructures
IOS: Interorganisational Systems
IT: Information Technology
JMLSG: Joint Money Laundering Steering Group
KYC: Know Your Customer
LACTEW: Los Angeles County Terrorism Early Warning System
LEA: Law Enforcement Agencies
MLRO: Money Laundering Reporting Officer

NBTC: National Border Targeting Centre
NCA: National Crime Agency
NPM: New Public Management
NSA: National Security Agency
NSL: National Security Letters
OFCOM: Office of Communications
OFWAT: Office of Water Services
OTA: Office of Technology Assessment
PNR: Passenger Name Record
PPP: Public-Private Partnership
SAR: Suspicious Activity Report
SOCA: Serious Organised Crime Agency
SWIFT: Society for Worldwide Interbank Financial Telecommunication
UK: United Kingdom
UKBA: United Kingdom Border Agency
US: United States
USA: United States of America

List of tables, figures and images

CHAPTER ONE

Consumer data and the war on terror

The new political economy of security and surveillance

The last decade has witnessed a blurring of the boundaries between public and private sector organisations in relation to national security. This blurring of boundaries has expanded from the private provision of physical security services and infrastructure, to the provision of the very data which enable decisions about risk and deployment to be made. Pre-empting the moves of risky, targeted individuals using vast datasets gleaned from any number of sources is *de rigueur* in neoliberal government discourse and doctrine. In this new politics of pre-emption, mined data about the past transactions and activities of citizens become the template for risk analysis about future threats (DeGoede, 2008) and produces a contemporary world where new securitised data flows between the private sector and government are forged. A political focus on the prevention of terror, catching suspects before threats materialise and denying suspected individuals access to material and financial resources, mobility and communications has driven security policy developments in the last 15 years, both in the USA, Europe and Australia. Marieke DeGoede (2008:163) argues that this move towards pre-emption:

> ... has the consequence of creating an extra-legal field of intervention in which administrative bureaucracies, immigration officials, consultants and financial dataminers are authorized to make sovereign decisions concerning the normality and abnormality of particular persons, behaviours and transactions, and to detain, question, monitor and freeze those considered abnormal.

Information about financial transactions, locations, and communications of citizens is 'among the most important and valuable' for national security.[1] This book examines the implications of this policy move to pre-emption for the private sector organisations that are mandated to provide such information: those who are empowered to make these sovereign decisions. How do these new government demands for information intertwine with the activities of private sector organisations, as their systems, processes, customers and employees are integrated into national security frameworks? National security processes are becoming diffused, decentralised and embedded into the consumption and employment processes which enable everyday life to function (Huysmans, 2011). How do private organisations achieve compliance with demands for customer data, how are their business operations altered and how are customer relationships and employees affected?

1 US Director of National Intelligence, James R. Clapper, statement Thursday May 31st 2013.

National security/public-private

Now running into its second post 9/11 decade, the expansive reach of the so called 'War on Terror' has reached a new level of maturity. Much of what we discuss in this book and indeed much of what has interested surveillance commentators, is the increasing prevalence and intensity of surveillance regimes on civilian populations or in 'the homeland'. There are numerous examples of such intensification: at the more obvious sites such as airports (Adey, 2009; Wagenaar and Boersma, 2012), and mega-events (Bennett and Haggerty, 2011; Giulianotti and Klauser, 2010; Fussey and Coaffee, 2011), particularly the use of drones (Surveillance Studies Network, 2006; Finn and Wright, 2012; Graham, 2012) to shopping malls (Koskela, 2000) and train stations (Adey et al., 2013). Moreover, the financial costs of mega-events and the rewards received by a coterie of technological firms such as Siemens and Cisco, raise further questions: are these measures warranted, what is their legitimacy and what of the profiteering that results for security businesses (Samatas, 2005)?

Collaboration between large defence contractors and the government has been a common feature of these developments. In the UK, government and the private sector have worked together on a number of fronts to intensify surveillance of the population. Recent, noteworthy collaborations have been between the DVLA and Experian in order to check on the background of drivers who make licence applications (Surveillance Studies Network, 2010) and between the various national governments and numerous large security companies in the securitising of mega-events – such as the Olympics (Bennett and Haggerty, 2011). Public-private collaboration was also an outcome of the many statutes which have been made law since 9/11. Britain's 'special relationship' with the USA in international security matters has had no small part to play here, extending, as it does, back to World War II. After George W. Bush called for 'Total Information Awareness' in 2001, and amidst pressures from the international community, it is perhaps no co-incidence that the New Labour government enacted a series of laws – e.g. the Anti-Terrorism, Crime and Security Act 2001, the Proceeds of Crime Act 2002, the Identity Card Act 2006 and the Immigration, Asylum and Nationality Act 2006, the Counter Terrorism Act 2008 – which mandated the mass collection of communications, financial, identification and travel data from the general population respectively.

Although some of these laws have fallen by the wayside, notably the National Identity Card scheme, which was abandoned in 2011,[2] a draft of new anti terror laws have been enacted or are proposed. It is the latest in a string of measures, which involve the private sector in national security. DRIPA (2014) reinstates the defunct Communications Data Bill and extends the UK government's communication surveillance powers internationally. The proposed legislation is premised on the private sector gathering, storing and transferring data to government when required. Internet service providers have to store communications data in the UK – such as the time, duration, originator and recipient of a communication and the location of the device from which it was made – for one year. Web browsing history and details of activity on social media, e-mail, voice calls over the Internet and gaming, in addition to phone calls would also be stored. Police could then access the details of the communication if investigating a crime, but would have to get a warrant from the Home Secretary to see the content of messages. One of the criticisms of the Bill raised by Liberty, a UK-based civil rights campaign organisation, were the potential dangers of outsourcing the monitoring of citizens to private companies.

Furthermore the instantiation of a Data Analysis Warehouse in Wythenshawe, near Manchester, which is fed by the eBorders regime legislated for in 2006, mirrors the function of 'data fusion centers' in the USA which were introduced in 1996. The Los Angeles County Terrorism Early Warning Centre (LACTEW) was the first (German and Stanley, 2008) many more have been formed in the post 9/11 era (Monahan and Palmer 2009; Monahan 2011). The centres were ostensibly designed to gather information from an array of sources and 'join the dots'. Data sharing information sourced from state and local police services, other emergency services, intelligence agencies, as well as private companies (Monahan and Palmer 2009: 618). The intention of the centres is to focus on not only terror, but also crime, hazards, trafficking, illegal immigration and all other 'threats'. The centralizing of information in preventing and dealing with threats is the over-riding incentive of the centres. However, while the centres are mainly located in police buildings, analysts are often external contractors with, as Monahan (2011) argues, less ethical conviction or indeed training. Private and sensitive in-

2 https://www.gov.uk/government/news/national-identity-register-destroyed-as-government-consigns-id-card-scheme-to-history accessed 11th December 2013.

formation available to contractors can include credit reports and banking details. Further tools of extracting information include National Security Letters (NSL), which allows governmental agencies to secure personal information from internet service providers, telecommunications companies, credit agencies or banks (Richards, 2013). It is this national security-based collaboration with non-defence industry partners – corporations whose products and services facilitate everyday life – which is the significant point of departure.

It would be naïve to assume that the US government provides the sole impetus for the collection of data from the private sector. The EU and Britain in particular have been willing partners in these developments (DeGoede, 2008). The USA, for example, were monitoring international bank transactions recorded in the Belgian-based SWIFT system illegally from 2001–2006. Despite initial outrage when this was discovered, Europe eventually gave in to pressure from the USA and produced the 'SWIFT agreement' in 2009, which gave the USA legal access to European financial transactions. The same is true of passenger name record sharing: a passenger name record, or PNR, is an airline's reservation system record which contains information about the passenger's booking. Mode of payment, seat and meal preferences would feature in the record. In 2011, Europe and the USA agreed to share PNR data, subject to a data protection safe harbour agreement. At the time of writing, however, the European Parliament has voted to suspend all security-related data sharing following the Edward Snowden revelations about the NSA and pending further data protection assurances.

Within a legislative and political climate which is constantly emphasising private sector data gathering for government, a new political economy of security and surveillance has emerged. An entire industry has sprung up pedalling financial surveillance software and various countries around the world – most notably South Africa[3] – are seeking to emulate Europe's stance on PNR and Britain's approach to passport data capture. However, as this new political economy emerges, national security also becomes the latest domain for the titanic clash of public and private interests to play out. As Sennett (2006: 18) argues:

3 http://appablog.wordpress.com/2014/03/27/gemalto-launches-integrated-border-and-visa-management-solution-for-fast-reliable-and-secure-immigration-control/ accessed April 2014.

> ... the multinational corporation used to be intertwined with the politics of the nation state ... the global corporation has investors and shareholders throughout the world and a structure of ownership too complex to serve national interests.

Even though the global corporation is seen by some as vastly more powerful than the nation state, Sennett asserts that the current state of affairs is inherently destabilising for the corporation. Diffuse international ownership structures have resulted in what Bennett Harrison terms 'impatient capital' – empowered investors who want short term results. The demands of impatient capital mean that corporations have to adapt to such external demands for short term profitability and modify their internal structures accordingly. The ongoing impact of such modifications – and hence instability – on employees and local economies persistently foregrounds the question of government regulation and corporate social responsibility. Recent scandals over corporation tax payments by multinationals, irresponsible lending by banks and, of course, the financial crisis, alongside the perennial questions of globalisation and the environmental impact of capitalist production, highlight how at odds these two vast interest groups can be. Under what conditions then, and with what effects, could large corporations be enmeshed into national security arrangements?

This research was conceived in 2008 and executed between 2009–2012. At that point there had been no worldwide scandal about the mass surveillance of communications data by governments and no wider question of the private sector relationship in their sharing of consumer data. There was, however, widespread public awareness of The Surveillance Society.[4] In Britain surveillance had become an election issue as every political party promised to ensure that the records of innocent people were removed from the National DNA database.[5] There was also some academic concern about the levels of security to be implemented during the 2012 Olympic Games[6] as well as about the locus of surveillance, given that private sector actors were becoming more involved in the delivery of government services (particularly in the Criminal Justice, Health and Welfare arenas). By the

4 http://news.bbc.co.uk/1/hi/uk/6108496.stm accessed 11th December 2013.

5 http://news.bbc.co.uk/1/hi/uk_politics/8515961.stm accessed 11th December 2013.

6 http://www.theguardian.com/sport/2012/mar/12/london-olympics-security-lockdown-london accessed 11th December 2013.

time we came to write up the research, however, the ground had shifted to foreground the controversial involvement of the private sector in national security surveillance.

The most significant development for our purposes was an open letter from the internet giants Google, Facebook, Twitter, Yahoo, Microsoft and LinkedIn imploring governments around the world to limit their surveillance powers in the name of free expression and privacy. At the time of writing, the letter can be found at www.reformgovernmentsurveillance.com. Nearly all of these corporations, with the exception of LinkedIn, were implicated in the 2013 Snowden revelations. Using the Protect America Act 2007 and the FISA Amendments Act 2008, and suggesting a security-commerce 'win-win', the NSA offered selected internet corporations immunity from prosecution if they 'voluntarily co-operated' with intelligence collection under its Prism programme. The NSA proceeded to collect customer data directly from the servers of Microsoft, Google and Yahoo (which provided 98% of the total dataset) as well as Skype, AOL, Facebook, YouTube, Apple and PalTalk. As they had been subject to court orders, the companies publicly denied any *voluntary* involvement with the programme, pointing out that they did not routinely share information with the NSA. Others gave the US government a less easy ride. Lavabit, Edward Snowden's e-mail service provider, chose to suspend its operations rather than provide the US government with the encryption keys for the information on their servers, which would have compromised the communications privacy of over 400,000 customers: a security-commerce lose-lose. Similarly, Twitter fought bitterly to prevent New York City gaining access to three months' worth of tweets belonging to Occupy protestor Malcolm Harris. Eventually they capitulated under threat of a contempt of court ruling and heavy fines.

The authors of the open letter would no doubt argue that communications data were gathered under legal duress (Bruce Schneier (2014) offers greater insight into the subtleties of such denials).[7] But a closer examination of the letter and its comments about trust on the Internet and the free flow of information in a 21st century economy belie the deep commercial concerns at the root of this protest. Threats to data security and privacy undermine brand value and commercial revenues as, in the words of Brad

7 http://www.theatlantic.com/technology/archive/2014/03/don-t-listen-to-google-and-facebook-the-public-private-surveillance-partnership-is-still-going-strong/284612/ accessed March 2014.

Smith of Microsoft, 'people won't use technology they don't trust'.[8] The government's initiatives to collect communications data for security purposes is conflicting with these companies' use of exactly the same data for commercial purposes. Indeed, this is at the very heart of their business models. For these companies, the creation of this securitised data flow is at odds with commercial priorities to the extent that they feel a need to protest against it. Their protest emphasised their need to maintain market position as trustworthy service providers they needed to appeal to the human rights agenda.

The contribution of this book is to identify the different ways in which a situation like this can be examined as an organisational problem. We show how it can have deep and multi-layered impacts on firms and their employees. These impacts stem from organisations having to respond to seemingly irreconcilable regulatory and commercial demands. They reverberate throughout stakeholder networks and detrimentally affect those with the least power. At the level of management practice, such demands can be seen as a regulatory compliance problem, a Corporate Social Responsibility problem, a strategic problem, a problem of market positioning and customer relationship management and finally, because of firms' internal politics of production, it is particularly a problem for labour. It is all of these things. We submit, however, that the unique experiential quality of these regimes for firms and their employees stems from the fact they constitute firms as intermediaries in large scale surveillance infrastructures. Firms are so constituted through practices of securitisation. Because consumer data is of such interest for national security, it grows in importance as a source of intelligence about the identities and activities of criminals and terrorists. As Huysmans (2014) argues, security is accomplished by foregrounding the insecurities inherent in any situation and, then, by mobilising resources to address them. Such insecurities stem from the legal responsibilities that come with identifying and handling potentially important information which is also commercially important. Introducing national security into commercial processes introduces new vulnerabilities into commercial operations. To construct something as a security matter does something, both analytically and practically, and it is this: To securitise a practice enacts it as dangerous.

8 See footnote 7.

Our empirical focus is on the Anti-Money Laundering/Counter Terrorism Financing (AML/CTF) and eBorders regimes. In the former scheme, banks, building societies, insurance companies and other financial services organisations are expected to look into the backgrounds of their customers for any suspicious activity when they buy a new product, and to monitor on-going transactions for anything out of the ordinary. If they have sufficient suspicion they are to submit a suspicious activity report to the National Crime Agency (NCA) (which used to be called the Serious Organised Crime Agency or SOCA), and are criminalised if they fail to do so. In the latter, airlines and their downstream supply chains are to collect passport data in advance of travel and transfer it to the UK Border Agency for screening against watch-lists, once again under threat of criminalisation.

As regulation put in place to feed into national security strategy, both AML/CTF and eBorders enact and constitute the activities which pervade the commercial practices of enabling people to travel and enabling people to use money as being shot through with *insecurities*. These new insecurities result in organisational tensions, competing pressures on resources as well as new anxieties introduced into the firms' operation. We find that many of these anxieties arise from the responsibilities which the regulations place upon organisations. Critically, as security practices become part of organisational routines, the ongoing burden of compliance rests with those in customer-facing jobs. We explore how national security processes interact with local workplace politics as they simultaneously intensify employee workloads and fundamentally change the roles they perform. Ultimately we counter recent arguments made by security studies specialists which suggest that the diffusion of security practices de-politicises security. As security becomes diffused it re-emerges in local political circulations, takes on a new political significance and opens up new possibilities for debate and resistance. We shall continue to see a struggle for legitimacy between the insecurities of security and business priorities as a securitised information flow gets forged.

Security, surveillance, transdisciplinarity

The book is positioned at the interstices of a number of business and social science disciplines – organisation studies; marketing; information systems and surveillance studies – and employs theory and method from all of them. Within this work, we draw on a generic definition of surveillance as *'any collection and processing of personal data, whether identifiable or not, for the purposes of influencing or managing those whose data have been garnered'* (Lyon, 2001:2). The argument we build is grounded not only in the theory, politics and practice of mass consumer surveillance and the discrimination, exclusion and 'social sorting' that results, but also in the critical observation that such practices are a fundamental part of modern organising. As bureaucracies began to form in order to manage local populations and to supply goods and services, so did information systems containing records of individual customers and citizens (Beniger, 1986). With the growth of information processing capacity, oversight, management and manipulation of these datasets became possible and modern surveillance was born. So, in our view, surveillance is not a 'malign plot hatched by evil powers' (Surveillance Studies Network, 2006), but a process of information gathering, analysis and application which makes modern living possible. Developing surveillance capacity is hence a normal part of both the governmental and private sector agenda: however, it is in the context of national security that these two surveillant domains begin to join up – and fragment – in new and powerful ways.

For private sector organisations, surveillance emerges as a way of monitoring resource use, productive capacity and managing risk and liability. A key development in the last 20 years has been that of customer relationship management (CRM) where customer transaction data are monitored to identify tastes, preferences and maintain the long term customer relationship. This strategy involves the mining of customer information to anticipate and service customer needs. It is used with varying degrees of sophistication across a variety of industrial sectors. It has proved particularly effective for many large financial service organisations, enabling them to identify lucrative groups of consumers, target products appropriate to those consumers and generate revenue (see Dibb and Meadows, 2004). Data analysis in the national security context often relies on the very same transactional data as CRM, and on similar statistical techniques, to sort through those data and identify suspicious activity. CRM focuses on the attractive customers, whereas national security practices seek to find the risky ones. Indeed the point of customer contact for the organisation is frequently the point of data capture for the national security regimes we investigate. Work on the dynamics of consumer surveillance which necessarily takes a critical approach to CRM (e.g., Andrejevic, 2009; Gandy, 2009, 2010; Turow, 2006, 2012) – has been extremely useful in highlighting the impact of surveillance activities on consumers, particularly the distributive justice implications of market segmentation, or 'social sorting'. Studies of customer relationship management acknowledge the likelihood for differential consumer treatment (Boulding et al., 2005), but see this as a problem for marketers rather than for society. But what of the organisations themselves? How do they walk the line between screening customers for attractiveness and for national security risk? An added complication is that the private sector response to any kind of regulatory scheme, let alone national security regimes, is under-researched from a regulatory compliance perspective. Compliance has been examined from a top-down, prescriptive and managerialist, perspective and has summarised the difference in organisational responses to regulation as 'idiosyncratic'. Internal organisational impacts have not been examined in any depth (Bennear, 2006; Kagan, 2006) although significant and relevant work emerged from the Organisation Studies literature in the 1990s (Knights and Murray, 1994; Knights and Morgan, 1997). Parker and Nielsen (2009) note the need for more empirical research which, first, examines how compliance is constructed locally and, second, which understands the impacts and effects of regulation.

We advance a transdisciplinary approach to this research problem. In the chapters which follow we use extant work on information infrastructures (Bowker and Star, 1999; Lyon, 2009; Sahay et al., 2009), stakeholder theory (Friedman and Miles, 2002; Mitchell et al., 1997) securitisation (Huysmans, 2011), labour process theory (Burawoy, 1985) and remediation (Bolter and Grusin, 2000) to frame the issue. We establish that the operationalisation of national security-driven data collection from the private sector relies on the successful alignment of consumer data, information systems, organisations and their employees to create a securitised information flow between the private sector and government. We reveal that the public and private interests that constitute and legitimise that flow are difficult to align. For firms, their information systems were the means of business; for the UK government, they are the means of addressing national security issues. Tensions caused by inadequate infrastructures, changing political landscapes and financial pressures all exert substantial effects on how this alignment is achieved.

As a result, new local activities arise which point back not just to local working conditions, politics and meaning systems within the organisation but also to the pan-governmental infrastructures that bring national security regimes into being. Resources and labour are redistributed, and established ways of working are disrupted. Critically, while system's investment underpins a lot of compliance activity, everyday adaptations and ongoing compliance is achieved by augmenting the roles of frontline service workers. Close inspection reveals how customer-facing workers juggle the competing imperatives of performance levels and a national security regime, with little room to negotiate the terms of either. In our view these national security regimes are an articulation of a neo-liberal political economy of surveillance which places responsibility on not only organisations but, most critically, their frontline staff. These are the people who face customers and deal with their varied and difficult demands and, ironically, who are under the most surveillance themselves. It is these people who are carrying the burden of national security work on a daily basis, as their jobs become extended, their work intensified, and who labour emotionally to reconcile the conflicting priorities of state and employer. Furthermore, as organisations work to align their competitive interests, internal processes, information infrastructures and employment patterns with national security regimes, there is a concomitant diffusion of security practices into a firms' internal politics of production. As security practices diffuse they are accompanied by notions of threat: not just the threat of terrorist attack, but the non-negotiable threat of criminalised fines for non-compliance. Here, we draw on the work of Burawoy (1985) and of Huysmans (2011) to argue that the securitised politics of exception re-emerges in the politics of production. By this we mean that frontline workers as well as their bosses feel little choice in their compliance with regulations but are compelled to comply for primarily economic reasons (Burawoy, 1985). Just as the Internet giants involved in the Prism scandal were granted immunity from prosecution for sharing data – thus preserving their reputation and financial position – the commercial imperative drives compliance in the firms we investigate as well. This does not, however, necessarily mean that commercial priorities and national security surveillance are easy bedfellows. The public interest of national security becomes subsumed under the market logics of commercial competition but it does not go down without a fight. Market logics themselves are not singularities, they are messy combinations of practices, technologies, resources and human beings which orient organising processes towards the priorities of capital accumulation and the pursuit of profit in the relevant marketplace.

Public-private collaboration: An unexplored mid-range?

The nascent realities of public–private collaboration for national security purposes should be viewed with growing interest by surveillance scholars and the wider academic business community for its theoretical importance as well as practical relevance. How surveillance practices become mobilised and operationalised are a central concern here. The eclectic and wide-ranging commentaries, such as the many works of Lyon (1994, 2001) and Gandy (1993, 2010), concentrate on system or society-wide analyses, building from historical and contemporary examples rather than examining practice. A growing number of empirical pieces have documented the local practices of surveillance in great detail. The Goffman-inspired microsociologies of Norris and Armstrong (1999), Norris and McCahill (2006), the empirical depth and detail provided by Smith (2007) and Neyland (2007) illuminate the local dynamics of surveillance and the production of some of its more problematic, discriminatory consequences (see also Ball, 2002, 2009). Whilst the presence of a macro level surveillant impulse and local consequences of surveillance are not called into question, the nature of the connection between the two has not been theorised in surveillance studies in a thoroughgoing way. Indeed Latour (2005) goes so far as to say connection between local clusters of surveillance activity – or Oligoptica – is highly unlikely to occur. However some of the criminological literature (including literature on CCTV) has examined how government-initiated 'responsibilisation' strategies (Safer Cities, Crime and Disorder Act, Crime Reduction Initiative, Challenge Competitions etc.) facilitated or created statutory requirements for public–private partnerships which, in turn, led to private-sector involvement in crime control (e.g. Gilling and Shuller, 2007; Skinns, 2008). However, the organisational arrangements were not examined in any depth in these studies. At a theoretical level, a number of potential theories exist which might address this shortcoming, such as Ball's (2002) 'Elements of Surveillance' or Haggerty and Ericsson's (2000)

'Surveillant Assemblage'. We explore these elements in depth and what brings them together. What are often missed are the tensions and dynamics in the interactions, as well as the heterogeneous meanings, interpretations and problems that produce surveillance practices and the systems that structure them. We term the connection that occurs in this 'mid-range' where surveillance practices are produced 'surveillance-in-action'. Essentially, we ask how do surveillance practices become real or lived? How, for instance, does a call centre employee interpret and act upon the Counter Terrorism Act 2008 when dealing with a customer they may find suspicious? A travel agent collecting passport information from a customer or a bank conducting 'know your customer' checks are not unusual procedures. They are considered by those who work in the travel and financial sectors as standard practices which take place because of regulation. But these are practices that we would argue constitute surveillance-in-action, and we would like to question how they become taken-for-granted parts of working life.

Notable studies such as Gilliom's 'Overseers of the Poor' (2001), Genosko and Thompson's 'Punched Drunk' (2009) and Ericson and Haggerty's 'Policing the Risk Society' (1997) do much to uncover the processes which produce surveillance in different institutional settings. Indeed Bennett (2001) has outlined that failures as well as positive strategic intentions are some of the driving forces behind the embedding of surveillance practices in everyday organisational worlds. Analysis to date has focused on documenting the discriminatory consequences and local practices of surveillance, or has focused on analysing the macro level surveillant impulse. The results of this focus are calls from a number of academic positions to render surveillance systems more accountable and transparent – in other words, to foreground surveillance-in-action – given their discriminatory consequences (PATS project 2011; Surveillance Studies Network, 2006).

Researching surveillance-in-action in the business context

Understanding how these regimes impact organisations from a Business and Management Studies point of view is a complex problem. An organisation is not a singularity. It is made up of different functions and numerous employees at different levels and locations. It has proprietary technologies, information systems and practices to help it achieve its goals. It has many internal systems of politics, power and social meaning. It also has external relationships that are essential for its existence. A regulatory requirement which demands consumer data to be passed out of the organisation to a regulator will have implications for these and other dimensions of the organisation.

Our analysis takes information flow as its starting point. Implicit within the definition of surveillance that we outlined earlier is the notion of an information flow. The gathering of data by someone or something in the position of authority to inform future actions to be taken in respect of those whose data have been gathered and analysed, implies the movement and flow of information. As a result, we approached this research problem by addressing how new securitised customer information flowed out of the organisation. We advance the idea that understanding the form surveillance takes in its mid-range involves understanding how the information flows which constitute it emerge. Taking a processual view of the organisation, involves aligning the organisational resources, actors, processes and stakeholders which produce these flows. Indeed, producing them involves the organisation and its employees overcoming some significant challenges posed by the prospect of creating information flows to government. The surveillant mid-range that we discuss above therefore encompasses those actors, other entities and processes which allow information to flow from the surveilled subject – in this case, the financial services or travel consumer – to the watching authority.

Conclusion

The government climate which features the diffusion of responsibility for national security from central government, through its agencies and then out into the private sector is here to stay. Whilst it may be controversial in legal and rights discourse to diffuse sovereign power into the realm of private capital, this process does enable a new security politics to emerge. Issues of security and compliance re-emerge in local politics of production in these private sector companies, as they become subject to the tensions and pressures of market competition in a world of scarce resources. In what follows, we explore these issues in more detail, both theoretically and empirically. Chapter 2 details the two UK sectors that are our focus and locates the debate within an array of literatures. First we discuss the financial services sector and the Anti-Money Laundering/Counter Terror Finance regulations. Second we discuss the air travel sector and the eBorders regulations. Chapter 3 maps the different interested parties, or stakeholders, in each sector and sets out their involvement. Chapter 4 draws on the information infrastructures literature to examine how the requirement for an information flow from the customer to either the UKBA or NCA prompted tensions in stakeholder relationships. Here we outline the broad competitive pressures which these regimes are subjected to. Chapters 5 and 6 adopt a strategic level of analysis in relation to the data. Chapter 5 examines the response of the UK retail travel sector over time and shows how commercial strategies adapt to incorporate the regulatory requirements. Chapter 6 examines the UK financial services sector and unpacks the relationship between customer relationship management and anti-money laundering practice. In both of these chapters we show how the market logics circulating in private sector firms shape and subordinate security issues. We show how front-line staff are framed as the group who had to reconcile security and commercial interests in an ongoing way. Chapters 7 and 8 examine this in more detail, discussing the impact of eBorders and AML/CTF on the frontline worker. Developing a new concept – 'remediation work' – the chapters show exactly how the regulations impact those in customer-facing roles and how they emerge in local production politics. Finally we consider the work as a whole and reflect on our experiences of doing this interdisciplinary project.

CHAPTER TWO

Market logics and regulation

Theorising private sector involvement in national security surveillance

Introduction

In this chapter we outline the characteristics of the different schemes and unpack how we theorised their effects on participating organisations. Each of the schemes investigated – Anti-Money Laundering/Counter Terror Finance regulation and eBorders – used very different means to extend the reach of national security into the private sector. But they have one thing in common: Their lifeblood is customer data held by the private sector. Each has attempted to create a securitised information flow about the everyday consumer from the organisations involved. In creating this information flow, the government's interest in maintaining national security has become entangled in the market logics of the private sector organisation. How this process unfolds, how the public interest logic of security becomes subsumed into the market driven logics of capital and how the private sector organisation is reconfigured by this process is the subject matter of this book.

Conceptualising something like this is not an easy thing to do. Perspectives from our disciplines – organisation studies, marketing, strategy, information systems and human geography – only ever seem to capture part of the story. A single perspective is not available, so in this chapter we draw creatively on what we feel are relevant and sufficiently powerful ideas to examine the entanglement and (mis)alignment of capital and government to the ends of national security. We conclude by identifying the elements of the organisation required to re-align with the schemes. We begin this chapter by discussing the two regimes, their origins, form, function and published research which has explored their dynamics and effects. We then continue to explore the theoretical frameworks that help to focus the work.

The regimes: eBorders and Anti-Money Laundering/Counter Terror Finance

eBorders

The story of the eBorders programme is complex. This is largely down to the programme's infrastructure and the politics of its emergence. eBorders stipulates that all travel carriers collect and electronically transmit passport information to the (now dissolved) UK Border Agency for all individuals travelling to and from the United Kingdom. Information must be transferred to the UKBA's 'National Border Targeting Centre' (NBTC) between 24 hours and 30 minutes before travel, which then uses the data to identify potentially risky individuals. Data are checked against watch lists and, on high-risk routes, travel patterns are subject to algorithmic surveillance to identify individuals suspected of being involved in dangerous activities. When concerns arise, border staff, including police, immigration and customs officers, are alerted. All information is then held for five years in an active database, and another five years in an archive with access on a strictly case-by-case basis. Although, currently, only air carriers are required to do so, eventually all sea and rail carriers must comply (Home Affairs Committee, 2009). Compliance is policed with a system of criminal fines. Airlines, travel agents, tour operators and seat brokers are affected by eBorders. They have had to implement systems and processes which enable passport data to be collected from customers and then transferred along the supply chain to the UKBA.

Although little research has focused on the implications of eBorders, its implementation was littered with controversy, and questions remain about its future at the time of writing. eBorders was conceived in 2003 after several governmental initiatives to update and improve the UK's system of electronic border control. In November 2004, UKBA launched a pilot scheme called Project Semaphore on a number of routes into the UK. Following what UKBA considered the success of this pilot scheme, and after a Regulatory Impact Assessment in 2005, the Immigration, Asylum and Nationality Act 2006 gave statutory authority for the scheme in its current form and the eBorders 'roll out' began in 2009.

Substantial challenges to the programme have arisen since then, some of which are of a legal nature. In June 2009, industry representatives were invited to state their concerns to the Home Affairs Select Committee. Commercial airlines from across the sector – national legacy carriers such as British Airways, charter operators (e.g. Tui or Monarch) and low cost airlines (e.g. Easyjet) – each voiced their opposition because of the cost implications and its impact on their business models. Tui Travel's evidence argued that 'for UK charter carriers, eBorders represents a £13 million a year cost to the industry against a £2,000 a year saving in not giving out boarding cards (Reals, 2008, p 2). Similarly, the Board of Airline Representatives UK (BAR UK) stressed that airlines needed to spend £450 million over the first 10 years of eBorders in order to make their internal systems compliant. The most pressing issue, however, has been the European Commission's warning that eBorders compromises European citizens' rights to freedom of movement (Whitehead, 2009). Belgium, France and Germany have also raised data protection concerns about the scheme. Although the European Commission, the UK's Information Commissioner's Office and the Home Affairs Select Committee state European citizens have the right to opt out of providing data, the UKBA insists that mandatory data collection is legal. The situation has not been resolved and rail and sea carriers are not yet investing in compliant systems. In the future, air carriers will have to absorb the enormous costs of amending systems to incorporate the opt-out. Due to these problems, eBorders has grown in political significance with debate featuring in successive Home Affairs Select Committee enquiries as the current government faces the growing criticism associated with its implementation.

Problems with the scheme began to arise following the appointment of the 'Trusted Borders' consortium: the team of companies charged with delivering the project. The impetus to create an electronic borders system was driven by US post 9/11 security demands, and the selection of the consortium was based on its experience with the US Department of Homeland Security (see Michaels, 2010). Furthermore the shape of eBorders was typical of New Labour's taste for all encompassing databases (Surveillance Studies Network, 2006). It emerged alongside the National Identity Register, the National DNA Database and the children's database Contact Point, each of which have been either abandoned or reformulated since the coalition government was elected in 2010. At the outset, the consortium consisted of defence contractor Raytheon (responsible for the overall management of the project), along with Serco (infrastructure), Detica (data analytics), Accenture (business change), Qinetiq (human factors), Capgemini (business processes) and Steria (system interfaces). In July 2010 the government sacked the main information technology contractor, Raytheon, because of significant delays in delivery (Ford, 2010; Kollewe, 2010). At the same time the Home Affairs Select Committee published a strongly worded critique which stated, 'the lessons learned from the pre-cursor Semaphore project had not been fed through to the contractors responsible for the eBorders Programme' (Home Affairs Committee 18.12.09 p. 9). Elsewhere in the legal press it was speculated that the UKBA had been less strict with the consortium than it had on previous IT projects and it emerged that relations between the UKBA and Trusted Borders were strained. IBM was then appointed; however their contract only ran until the London Olympics in 2012. Since then no further contractor has been employed (BBC, 2012).

Issues with infrastructure arose because of the design decisions made by Trusted Borders which disadvantaged parts of the air travel sector. Rather than design a centralised portal that all travel companies – airlines, tour operators and travel agents – could use to transfer passenger information, it decided to rely on airlines to collect passenger information and then pass it on, reducing the cost to government. Many of the larger national 'legacy' airlines use Global Distribution Systems (GDSs) to manage their bookings. GDSs are computer interfaces that directly handle the booking systems of numerous airlines so that agents can book air seats independently of any other travel product. A travel agent can search on the GDS for seats and fares on all airlines flying to a destination at a specified time and date. Following pressure from the Association of British Travel Agents, the three main GDS companies changed their systems to enable travel agents to enter passport information on booking, which is then transferred directly to the airline and on to the UKBA. This made it easy for legacy carriers to transfer passenger information at low cost and they easily met the eBorders requirements via the GDS. However, this service was not available to anyone selling retail travel products (see Home Affairs Committee, 2009, Evidence 2: 30). Legacy airlines effectively had their own centralised data collection point via the GDSs, but the retail operators did not. This is because retail travel companies tend to charter aircraft to carry their passengers: Charter flights are not listed on Global Distribution Systems (GDSs) as they are not available for independent sale. The retail sector had to make its own arrangements to facilitate the transfer of passenger information. New customer websites, self-service kiosks in airports and help lines needed to be set up at the retail operators' expense to collect data from customers and make them aware of the new requirements. The Association of British Travel Agents Airline Group even investigated the feasibility of purchasing a centralised data collection point for the retail sector, but no appropriate service was commercially available.

As such, significant investment in internal systems and training was required throughout the travel supply chain. A further problem was that the scheme dictated that airlines gather passport data separately from customers while they were overseas, creating new datasets for inbound flights, rather than rely on the data collected from the same passengers on outbound flights. This was unnecessary for retail operators as exactly the same passengers would travel on inbound and outbound flights as part of a package deal so their passport data were already with the UKBA. In addition to doubling the administrative burden, destination airports, which were often in less wealthy countries or were tiny islands and in some cases did not even have reliable electricity supplies, also had to invest in compliant systems. The cost of this investment was passed on to the airlines in increased landing fees. Since then, the Home Affairs Select Committee has highlighted the wasted investment made by the industry in relation to eBorders. Despite the travel sector capturing data in advance, passport data are still required to be captured at check-in. The information is being captured twice, thus removing any benefits of speed or efficiency that may have been offered to passengers at the airport (Home Affairs Select Committee, 16 July 2012). The instability of Trusted Borders and poor communication between the Travel industry, UKBA and the consortium have also promoted deep political wrangling; especially with a governmental spend, to date, estimated to be £750 million (£60 million over budget) and with the full roll-out of the programme yet to be ratified.

Anti-Money Laundering/Counter Terror Finance (AML/CTF)

The story of the current AML/CTF regime is less fraught with controversy than eBorders. 'Money laundering describes the process through which illicit profits are hidden from authorities, often by using a combination of complex financial transactions and financial secrecy, and re-introduced into the financial system under the guise of legitimate transactions' (Sica, 2000: 47). The rationale is that most criminal activity is believed to be financially motivated (Assets Recovery Agency, 2003). Money is not only key to executing crime, but also enables the 'training, recruitment, reconnaissance, and radicalization' of would be terrorists (DeGoede, 2012, p. 5). Hence, hindering the movement of criminal funds is widely believed to lead to a reduction in crime (Harvey, 2005). On the other hand, financial transactions leave traces – the electronic footprint – which can be pursued and connected to individuals and their networks (DeGoede, 2012), so financial intelligence has become a key component of modern national security (DeRosa, 2004). Governments around the world have adopted a range of initiatives to curtail criminal activity, on the basis that crime is deemed to cause social, environmental and economic harm (FATF, 2010). Such initiatives that focus on limiting the movement of money resulting from, or used to fund, criminal activity are referred to as 'Anti-Money Laundering and Counter Terrorism Financing' programs. Financial services are instrumental in the movement of money globally, affecting directly or indirectly all of the population of 'developed' countries (Leyshon and Thrift, 1999; Zdanowicz, 2004). Most banking transactions are now automatically recorded and easily traceable (Levi and Wall, 2004). These records are deemed to be a particularly reliable source of criminal intelligence, and they play an increasingly important role in law enforcement (Aufhauser, 2003; de Goede, 2012).

AML/CTF schemes have been in operation in the UK, in one form or another, since the late 1980s. The origins of AML lie in the US 'War on Drugs' in the 1980s (Gill and Taylor, 2004), and in the 1988 Vienna UN Convention it was agreed that states needed to criminalise the laundering of the proceeds of drug-related offences. In parallel the 1988 Basle committee on banking reform also identified the principles and purpose of anti-money laundering rules. An international regulatory body which established the principles of anti-money laundering practice, called the Financial Action Task Force (FATF) was created by the then G-7 heads of state at their Paris summit in 1989. These principles were taken up by national legislatures which empower law enforcement bodies, such as the UK's National Crime Agency, to act. In the UK, legislation relating to AML first appeared in the Criminal Justice Act of 1988. Since then the schemes have been revised and updated, extending the range of activity to be monitored, and requiring the deployment of systems and controls to combat money laundering and terrorism financing (FSA, 2003). Following the 9/11 attacks, the international Financial Action Task Force (FATF) renewed their guidance on Anti-Money Laundering to incorporate new guidance on Counter Terrorism Financing. Money laundering as an offence took on new significance: rather than being an adjunct to organised crime, it was seen as a means by which terrorists could further their causes and the stakes for participating organisations were raised. Following the introduction of the Proceeds of Crime Act in 2002, failing to disclose suspicious transactions in respect of all crime became a criminal offence (JMLSG, 2006). This applies to all employees of financial institutions.

In the UK, AML regulations require banks to collect and use customer information 'over and above the basic identification information' and to monitor how customers use the 'firm's products and services' (FSA, 2003). Very broadly, the recommendations, now embodied in legislation and industry guidance, stipulate that financial services institutions should 'know their customer' – in other words, conduct appropriate identity checks when an account is opened. Further, that they should perform 'customer due diligence' – in other words, to monitor transactions in an ongoing manner to identify anything suspicious. Banks have long collected and stored customer identity and transaction data, as a way of minimising exposure to bad debt, and identifying and responding to customer needs. The industry has been using numeric credit scoring systems since the 1960s (Capon, 1982), and has become one of the most advanced in terms of customer management (Ryals and Payne, 2001).

The Financial Services Authority provides guidance as to how institutions should respond to the regulations: each institution is to identify its most risky products, customers and territories and focus their AML efforts there. Each institution is also free to choose the technological solution that best meets their operational characteristics. All members of staff are required to report individuals or transactions that they view as anomalous to relevant staff within the organisation, namely the Money Laundering Reporting Officer (MLRO). Guidance as to what constitutes a suspicious transaction is widespread in the trade literature: if a customer is unusually interested in AML/CTF provision on opening an account; if an account has sudden and unexplained activity, including money transfers; if a customer has multiple accounts; if there are sudden inflows of funds which are beyond their known income or if a customer seems to provide false information: these are all grounds for suspicion (Longfellow, 2006). Customer-facing frontline staff are particularly affected as they deal with customers and their transactions directly. They are responsible for spotting suspicious activity and constantly have to reconcile the demands of sales and security (Amicelle, 2011). Financial crime specialists are also responsible for checking whether names on watch lists from either NCA or HM Treasury appear in their transaction records. They also run profiling analyses on all transactions to identify any unusual or suspicious activity based on recognised patterns of behaviour. After analysing those reports, and conducting further investigations, the organisation must then pass on information about suspicious activities to the National Crime Agency, which will investigate and take appropriate action (Backhouse, Canhoto et al. 2005).

The AML/CTF scheme has become embedded in the systems and practices of UK retail banks (Gill and Taylor, 2003), even though there is scepticism regarding the effectiveness of the initiative to detect and prevent crime (Webb, 2004). As the NCA watch lists have expanded, reporting levels have risen exponentially, from 5000 reports per year in 1995, to 278,665 in 2012.[9] A risk-based approach, which allows participating organisations to choose where they direct their AML/CTF efforts, initially resulted in a round of

9 http://www.nationalcrimeagency.gov.uk/publications/19-2012-sars-annual-report/file accessed 26th November 2013.

defensive reporting. Organisations were concerned about the reputational, commercial and criminal consequences of not reporting something which later turned out to be suspicious. The AML/CTF regulations have now been extended to solicitors, accountants, estate agents, money transfer operations and dealers in high value goods. Although most SARs are made by the big four banks – National Westminster Bank, Lloyds TSB, Barclays and HSBC – the widening of the legislation to incorporate smaller and more varied organisations has caused reporting standards to vary.

A small body of academic research has emerged which addresses its impact on participating firms. Many of these findings are critical of the regulations and highlight the deep impact of AML/CTF on banking practices. Sharman and Chaiken (2009) noted that firms felt the training requirements and the financial burden of the regulations to be high. Canhoto (2008) argued that AML/CTF compromised the fiduciary duty of financial organisations to their customer, and Ryder (2008) suggested that the scheme was heavily reliant on the goodwill of organisations as there was little to gain, but much to lose, from the regulations. Although the scheme assumes that criminal activities can be detected in financial data, illegitimate transaction patterns may differ little from legitimate ones (OTA 1995) and originate from many different kinds of criminal activity (Yeandle, Mainelli et al., 2005). The technical limitations arising from the lack of reliable profiles mean that financial intelligence is, in essence, a 'speculative endeavour' (de Goede, 2012, p. 58). Money laundering evolves as criminals experiment with novel strategies and instruments (Harvey, 2005). Also, reporting customers runs against the traditional strategic objectives of banks, the culture of privacy (Donaghy, 2002) between a bank and its customer and how performance is assessed and rewarded (Canhoto, 2008). Furthermore there are wider consequences of the regulations for society. Gill and Taylor (2004), for example, argued that the regulations created a financial underclass of those who were considered too risky for financial services products and created onerous problems for those who were deemed 'unusual' clients. Critical writers such as Amicelle and Favarel-Garrigues (2012) point out the difficulties Money Laundering Reporting Officers face in reconciling the dual interests of national security and commercial priorities. Furthermore they highlight that the embedding of national security decision making in private firms renders it opaque and unaccountable to the outsider.

Viewing the schemes together

When viewed together these two schemes have some important similarities. Both involve powerful global industrial sectors which interact with the general population as customers on a daily basis. Financial services and air travel are sectors which are seen by government to facilitate the means of criminality and terrorism, either by enabling the movement of money or the movement of people (Amoore and de Goede, 2008). Both schemes also see the customer as the main security threat because they assume that traces of risky individuals are lurking somewhere in the consumer data. Effectively customers are re-inscribed as security subjects by virtue of their data being collected for those purposes. Watch lists, profiling and behavioural analysis are used to analyse data in both schemes. Indeed each is imbued with a politics of pre-emption (de Goede, 2008), assuming that these forms of data analysis will identify and apprehend risky individuals before they commit harm. By extension the schemes transform business organisations into agents of the security state by turning them into data conduits for the security regimes. Firm participation in both of the schemes is mandatory: they are fined for non-participation and are governed by the threat of criminalisation for non-compliance. The flow of information within the scheme is from the organisation back to the state. Little feedback as to the quality or timeliness of reporting is received from the state by participating firms. It is also arguable that in both sectors customers are not fully aware of what is being done with their data, as firms do not always tell customers that their data are passed on to government to avoid damaging the customer relationship.

However there are some key differences between the regimes associated with the volumes of data that are transferred to government by firms in each sector and what firms are required to do internally in order to comply. One scheme, AML/CTF, is mature and well embedded while the other, eBorders, is newer and failing, so the aims and scope of the scheme keep changing in the latter but they are relatively stable in the former. The extent to which firms are required to adapt to the schemes' requirements thus varies greatly. Furthermore the schemes are differently configured, so participating firms are required to do different things. In AML/CTF, firms are required to analyse consumer data for suspicious activity before transferring it to NCA, whereas in eBorders data are transferred en masse where it is analysed by the UKBA. As such in AML/CTF firms assess the riskiness of transactions locally but firms do not perform any kind of risk assessment in eBorders. The two schemes also have different infrastructural requirements: AML relies on data transfer in very specific forms (a Suspicious Activity Report or SAR) on a one-off basis via a NCA web portal and eBorders relies on mass data transfer using airline legacy systems. Data volumes, legacy systems and sensitivity in financial services would prevent data being completely available to government in all cases. Customer account data is also very detailed and extensive, documenting everyday lives very closely as individuals spend money through the course of their daily activities. Mass transfer of such data would be unmanageable. Travel data only relates to customer identification and travel consumption which, for most individuals, only takes place once a year if at all. It is more difficult to glean a detailed picture of the customer and their intentions from travel data than it is from financial services data. The UKBA data warehouse matches travel data with data gleaned from other governmental data sources, such as the Police National Computer, Social Security data bases and so on. A further implication for firms is the nature and extent of the training requirement. In AML/CTF, detailed and ongoing training is required for all staff so that they understand the implications of the regulations within their roles and can recognise suspicious activity when it occurs. Training in the travel sector concerned the transactional aspects of collecting passport data from customers and inputting it to the databases of different airlines so that they could be transferred on time. Thus, while the same issues drive these two regimes and their implications for firms are similar, they differ in what they require firms to do because of the depth of information available about the customer, the types of data that are transferred and the way that risk is assessed.

Theorising organisational involvement

In Chapter 1, we outlined how critical it has now become to understand the implications of public–private blurring for surveillance purposes. It is clear that the surveillance practices embodied in these two schemes represent an exciting site of study which exposes not just the macro commitments of, but also the micro implications for, private sector organisations affected by national security programmes. In positioning organisations as agents of security by annexing their information gathering, analysis and transfer capacities, national security state surveillance pierces and reconfigures the internal workings of organisations. Although Huysmans (2011) has argued that the diffusion of security practices into the private sector has de-politicised security at state level, the resurfacing of security practices in the organisation may well precipitate a local re-politicisation as production processes are reconfigured by security (see also Aradau et al., 2008). Indeed, Louise Amoore (2009) has pointed out that the algorithmic surveillance which features in both regimes embeds warlike national security practices within every life, including, by extension, everyday organisational life. It is a technical, non-violent way of pursuing homeland security (de Goede, 2012). When we look to organisation theory, however, we find that indications as to what these reconfigurations might be are not clearly in evidence. A number of organisational literatures do provide some clues. In this section we examine literature concerning management-based regulation, public–private partnerships and private security outsourcing to set the theoretical context for the work. From these literatures it is possible to identify the organisational elements that will be implicated in the schemes and required to align with their objectives.

Management-based regulation

The AML/CTF and eBorders schemes are typical of what Hutter (2006) terms non-state regulation. She notes that the considerable moves to deregulation in many countries in the late 1980s have been followed by a marked move back to re-regulation because of the inclusion of non-state actors in regulatory processes. Parker and Nielsen (2009) outline how re-regulation has occurred. Rather than a return to a centralised, legalised regulation of industry, diffuse regulatory bodies have emerged which give businesses the responsibility to respond to public interest agendas through intensified reporting schemes accompanied by incentives. As part of a phenomenon which has been labelled as 'regulatory capitalism' (Levi-Faur, 2005; Braithwaite, 2008) public governments have been redesigned to work with private sector and civil society to promote social good as well as economic prosperity. This marks an extension of the neoliberal state through a number of regulatory bodies (e.g. OFCOM, OFWAT, FSA) which seek to shape the activities of, and regulate power within, strategically important business sectors, particularly those which are privatised state businesses. Braithwaite (2008) observes that rather than the state 'hollowing out', it has been 'filling out' with a move to regulation at all levels and more hybrid public–private forms of power and governance. And yet at the same time, he argues that markets, on a global scale, have been thriving in this climate. Critically, regulatory capitalism favours larger organisations as they have the capacity to absorb a regulatory burden. Growth in non-state regulation has been witnessed in a wide range of areas including environmental management, food standards and pharmaceuticals (Henson and Heasman, 1998; Westley and Vredenburg, 1991).

In considering the implications of non-state regulation for organisations, Coglianese and Nash (2006) assert that moving regulation to private sector companies will have operational implications for these companies, and will seek to 'penetrate and shape what goes on inside private sector firms' (p. 3). Information gathering, standard setting and behaviour modification are recognised as key components of regulatory control (Hood et al., 2001). That said, little has been written about how private sector organisations respond to government regulation locally, particularly muscular regulations which first require mandatory information, reporting to government by companies, and then purport to change how companies operate internally. Only a small body of work led by Cary Coglianese and Jennifer Nash (2006) has attempted to explore the dynamics of these kinds of regulation which primarily occur in the environmental field. The work examines the advantages and disadvantages to policymakers of relying on management to implement different forms of environmental regulation.

Using management-based strategies implies an element of policy which is directed at changing the behaviour of management towards the environment, either through direct regulation which features compliance and fines (quadrant 1), incentives (quadrant 2), implementation of supply chain standards such as ISO 14001 (quadrant 3) or the compulsory use of environmental management systems (quadrant 4). Pressures on management to comply can either come from government or industry partners. The authors categorise the different strategies using a 2x2 matrix:

	Government user	Non government user
Management required	1. Management-based regulations Sanctions and fines through legislation and codes	3. Management-based mandates e.g supply chain management using ISO 14001
Management encouraged	2. Management-based incentives Reward and recognition through awards schemes	4. Management-based pressures e.g. pressure to achieve ISO standards as an industry standard

Table 2.1. Management-based strategies (Coglianese and Nash, 2006: 14).

It is useful to consider our two schemes in the light of this simple matrix. They represent a curious combination of two management-based regulatory styles delineated above. It is clear that AML/CTF is pure management-based regulation, in which management (and staff) are required to comply with the regulation on threat of sanction based in legislation. In respect of eBorders, a curious combination of quadrants 1 and 3 emerges. For air carriers, persistent non-provision of passenger data before travel results in fines, indicating that it features management-based regulation. However as the eBorders legislation only applies to air carriers they have become responsible for collecting passenger data from downstream air seat vendors such as travel agents, tour operators, online vendors such as Expedia and seat brokers. The legislation has resulted in the travel supply chain developing ways in which data can be transferred to government in the right format. New infrastructures to share this information have been developed. Therefore, eBorders is a combination of management-based regulation and management-based mandate requiring significant infrastructural alignment.

Snyder (2006) argued that management-based regulation works best in three sets of circumstances: First, when the organisational population is diverse, so each organisation can choose their own response. Second, where the information collection burden on the regulatory body would be very high if it was not decentralised. Finally, where the risks associated with regulation are uncertain. According to this argument, the designs of AML/CTF and eBorders respectively make sense. The costs of data collection would be reduced for government in both cases. But, if we drew on this literature alone, we would still be none the wiser as to the implications of these schemes for the organisations involved beyond the fact that it somehow involves management action and organisational values, responses are idiosyncratic and infrastructural adaptations may be required. A critical aspect of the environmental regulation literature is that individual companies are depicted as having very individual responses. Kagan (2006), for example, studied 14 pulp mills in Australia to identify different types of response according to the strength of the company's environmental ethos, the degree of their responsiveness to environmental demands and the assiduousness of managers in implementing routines. He also highlighted how companies were reluctant to invest vast amounts of resources in compliance and would only do so when regulations where tightened. More generally, Porter (2011) considers the effect of regulatory transfer on the operation of private sector firms. He cautions that whilst regulators can determine the objectives or targets of regulation, they should not mandate the approach to meeting the regulation, which he sees as 'blocking innovation and almost always inflicting cost on companies' (p. 74). Non-state regulation involving private sector companies with divergent primary interests risks tensions and conflicts as elements of the organisation align with the new regulatory requirements (Coglianese and Nash, 2006; Porter, 2011). For the organisation theorist, these conclusions, while a useful point of departure, indicate the perspectival limitations of the literature, concerned as it is with policy choice as the primary unit of analysis.

Another way into this problematic would be to examine the literature on Corporate Social Responsibility (CSR), particularly that which concerns how and why organisations seek to pursue non-commercial goals. Regulatory compliance could be seen as part, but not all, of the activities which constitute CSR. The literature on CSR is vast and a comprehensive review would not be appropriate here, but we find Gerry Hanlon's work particularly informative in relation to our work. Hanlon (2008) acknowledges three bodies of scholarship which underpin CSR. The first stems from the field of business ethics and examines the different ethical justifications advanced by firms who purport to 'do' CSR. This section of the literature is criticised because it does not engage with the organisational politics which drive many CSR activities. The second is termed the 'business' perspective and evaluates CSR activities as it would any other aspect of commercial activity. This perspective typically holds that CSR is worthwhile if it does not damage shareholder value. Hanlon's third, more critical perspective is especially useful. This perspective (Hanlon 2008; Hanlon and Fleming 2009) explains how firms may react to regulation and take advantage of the new market opportunities which arise as a result. Hanlon uses Italian Marxist perspectives (Aglietta, 2000) to argue that new forms of accumulation based on flexible specialisation can explain firms' 'tactical forays into the social body'. Flexible specialisation explores how, in a post-Fordist world, firms exploit new markets and tailor offerings to them rather than providing a 'one size fits all' product as they did under Fordist capitalism. One of Hanlon's key arguments is that firms see CSR as an opportunity to marketise the non-commodified world, as a form of commodity fetishism (Marx, 1976). Therefore, a large multinational might enter into providing products or services which promote care for the elderly if it can generate new revenues providing such products and services, or may manipulate bio-diversity or fair trade to its advantage. The multinational will not enter into those potential markets for purely altruistic ends as some CSR scholars may argue.

It may well be the case that firms in the travel and financial services sectors make similar moves in relation to mandatory national security compliance. Critical perspectives on CSR suggest that affected firms may seek to exploit any commercial aspects of security driven consumer data collection in whatever way they can. Indeed, detailed ethnographies of regulatory compliance in banking undertaken in the 1990s seem to suggest that this is the case. Quack and Hildebrandt (1997) and Salomon (1999) conducted studies in the German and French banking sectors. Both sets of scholars conclude that institutional responsiveness to structural and regulatory changes was mediated through a set of internally derived 'market logics', linked to the firms existing business practices and its market position. To delve further into the operation of market logics in affected organisations, we turn to another literature on the public sector: public-private partnership and new public management.

Public-private partnerships and new public management

In many ways new public management (NPM) shares the same roots as non-state regulation in that it stems from the break-up of the large state owned enterprises of the 1970s and 1980s. NPM refers to how the management practices of the newly broken down public sector changes as a result of the organisations' exposure to market forces, particularly through the introduction of internal markets and competitive tendering for government services. Research which addresses the impact of new public management frameworks on public sector organisations indicate that there are adverse outcomes, disruptions and tensions for information systems, labour and value systems within participating organisations.

In an effort to promote better management of the public purse, NPM approaches scrutinise public sector managers' performance and that of their organisations much more closely than had been experienced in the old public sector. Performance indicators as well as a degree of accountability to external customers became the norm as many government services were outsourced through competitive tendering processes. Governmental control became centralised, however, as the performance of service providers was subject to exacting monitoring. The organisational arrangements which emerged are reminiscent of the issues raised by AML/CTF and eBorders in that the established boundaries between the public and private sectors became blurred. NPM academic John Benington (2007) noted that the forms of organisation which emerged at these blurred boundaries were somewhere between the hierarchical bureaucratic state and the competitive private market, characterised by adaptability, responsiveness and a focus on facilitating everyday life for citizens.

In the UK, the tendering schemes in place since the 1980s took on a number of different forms. These policy frameworks mirror, and in many ways legitimise, the outsourcing of national security activities to the private sector. Compulsory Competitive Tendering, introduced under Thatcher's government and now abolished, forced local authorities to tender for the provision of services, such as refuse collection, at set intervals, conforming to national guidelines. Under New Labour, the Best Value policy framework promoted partnership between local government and service providers rather than compulsory competition at regular intervals. The latest incarnation of the framework, Public–Private Partnership, involves private sector finance and management expertise and innovation being used to 'modernise' public services (Smith, 2012). In PPP, sections of government agencies are given over to private sector management teams to bring them up to date. Its critics point out that because many stakeholders are involved, and because power is distributed unequally between stakeholders, the interests of the least powerful – in other words, labour – will be sacrificed in the interests of profitability (Smith, 2012; Flecker and Meil, 2010). 'Hidden costs' relating to the impact on workers are in danger of being overlooked (Grimshaw et al., 2002). Empirical work which examines the impact of partnership programmes on working conditions confirms these fears. Under the 'Best Value' (BV) policy, Roper, James and Higgins (2005) found that staff and their unions benefitted the least in terms of morale, job satisfaction and pay and conditions. Taylor and Cooper (2008) reported appalling working conditions in a prison run under PPP, accompanied by a downward pressure on wages, high staff turnover and understaffing. Smith's (2012) study of the National Savings partnership with outsource.com also reported that work intensification and proscriptive performance indicators (where the new company was fined for failure rather than rewarded for success) put pressure on workers. Smith (2012) also observed IT problems stemming from the inadequate adaptation of legacy systems, which compounded the problem. Deskilling following job standardisation to ensure employment flexibility in shared service centres has also been observed (Howcroft and Richardson, 2012). While this literature outlines how the private outsourcing of government services is in danger of degrading work, an interesting finding concerns how the public sector service ethos remains intact in affected workers (Grugulis and Vincent, 2009; Hebson et al., 2003).

We should, however, note some key differences between NPM approaches and the regimes in question. While the regimes as a whole could be viewed as public-private partnerships, it is questionable whether the relationship between participating firms in each respective regime could be viewed in such a way at all. Involvement in the regimes is mandatory – there is no tendering or bidding process – and the organisations have little or no bargaining power to shape the outcome of the programmes. Furthermore, while PPP and BV concerned replacing government services with private sector expertise, the regimes supplement and extend the power of NCA and the UK Border Agency respectively as private sector organisations provide them with data. The regimes are governed by threat, through criminalisation and fines, rather than through varieties of KPI. Considering the research findings relating to the impact of NPM approaches have indicated some potential effects within the firm. Working conditions may be intensified or degraded and the public sector service ethos relating to security may well challenge the dominant private sector perspectives dominated by market logics in organisations affected by the regimes. Indeed where security functions are delivered by the private sector, this is exactly what has been observed to happen. We consider this literature in the next section.

Outsourcing security

The final area we consider as theoretical context is the outsourcing of state security functions to the private sector. The private security sector comprises a wide range of organisations which provide physical security services, technical advice and training in various military and civilian contexts. Its growth is attributed, in no small degree, to a number of pervasive national cultural, economic, political and social changes in recent years. Successive state fiscal crises, the under-resourcing of police forces, the rise of a neo-liberal mentality which seeks to responsiblise non-state actors for security, coupled with public worries about crime and terror fuelled by the media, have all contributed to the current state of affairs (Goold et al., 2010). Recent theorisations of the phenomenon have highlighted how, far from an old style Weberian monopoly of state power in relation to security (Gerth and Wright Mills, 1970), the security landscape now comprises a series of interlinked heterogeneous nodes through which governance is achieved (Johnson and Shearing, 2003). These nodes feature the aforementioned range of private security providers as well as government agencies. In spite of this heterogeneity, Loader and Walker (2010) highlight an enduring tension between the public good of maintaining security and private sector interests of profit-making. In what is termed 'The New Political Economy of Security', White (2011) outlines a dialectical relationship between these competing interests, arguing that firms need to internalise more public-spirited security values. There are clear dangers associated with the use of private security contractors where this has not occurred as they are not publicly accountable for their actions (Baker and Pattison, 2012). More generally, this tension is observed throughout empirical studies of security work. Nalla and Hwang (2006) observe constant tensions over competency between private and public police forces in South Korea, and VanCalster (2011) illustrates the difficulties shopkeepers have in incorporating government sanctioned anti-shoplifting responsibilities into their businesses in the Netherlands. The nature of security work itself is widely documented as having a reputational problem, among other things, as it is poorly paid, deprofessionalised and offers low levels of training and opportunities for collective organisation (Wakefield, 2003; Thumala et al., 2011). According to this literature, we might expect to find that the private sector 'nodes' which feature in AML/CTF and eBorders are characterised by tensions which arise from conflicting public and private interests and additional security duties to be unrewarding for workers, in both formal and informal ways. However, as with NPM there are also some key differences between the nature of the regimes and the private security sector which are worth noting. The security work which occurs in the private security sector and is discussed in its literature is that of security professionals: contractors in Iraq or Afghanistan, private police, security guards, technology vendors and trainers. Those affected by the regimes are not security professionals and, as such, the security work undertaken by individuals in participating organisations needs to be understood as a new element of their everyday tasks.

Conclusion: Re-politicising security in the politics of production

In this book our main contention is that, as security matters become decentralised throughout networks, they do not, as security scholars suggest, become depoliticised. We suggest that they re-emerge in local political economies to intensify existing political-economic relations, exploiting some and privileging others. For the private sector organisations involved in our study, we suggest that decentralised security practices become re-politicised as they embed within localised politics of production and as engagement in security processes becomes marketised. The implication here is that the relative distribution of the security-burden throughout the organisation mirrors existing resource distributions as well as distributions of advantage and disadvantage within the organisation. As security demands become embedded within organisational routines, so does the ongoing work of employees as they gather information which feeds this infrastructure. Security work then starts to become invisible, in the same way that emotional labour (Hochschild, 1983) and immaterial labour within the creative industries (Lazzarato, 2006) are invisible yet profoundly commercially exploited.

Hardt and Negri's (2000) discussion of immaterial labour in *Empire* can be extended to incorporate the distribution of security work as we observe it. They suggest that 'the communicative labour of industrial production that has newly become linked in informational networks, the interactive labour of symbolic analysis and problem solving, and the labour of the production and manipulation of effects is immaterial labour' (Hardt and Negri, 2000: 30). The security work undertaken by front-line employees, gathering data from customers without damaging the customer relationship is, as Hardt and Negri (2003) describe, the production and manipulation of an effect: maintaining the veneer of friendly service while producing security data.

In focusing on these new political economies of surveillance and security within organisations, we are not arguing that prior to introduction of these schemes the organisations involved were in some sort of stasis. In presenting this analysis we follow Van de Ven and Poole (2005) who view organisations in strongly processual terms (Chia, 2004) as well as Farjoun (2010) who views stability and change as a duality. Within this view, organisations comprise a set of organising processes. The organisation is produced by the continual construction and reinforcement of operating relationships, such as stakeholder relationships and infrastructural alignments, *as* organising processes. As Van de Ven and Poole (2005: 1380) argue:

> stability and change are explained in the same terms: stability is due to processes that maintain the organisation so that it can be reified as the same thing by some observer(s), while change occurs when the processes operate in a manner that is reified by observer(s) as changing the organisation. In both instances, stability or change are judgments, not real things, because the organisation is a process that is continuously being constituted and reconstituted.

This view is of particular relevance when considering information technology: a long history of technological research in organisations, typified by the work of Wanda Orlikowski (1992), proposes the view that technology and organisation continually reconstitute each other as new technologies become implemented. We were also inspired by Knights and Murray's (1994) classic processual and political study of information technology change. Knights and Murray (1994) argue that technological change is an inherently political process which is brought about by the interaction of numerous 'conditions of possibility'. Such conditions interweave organisational structural contingencies with individual identities, macro socio-economic conditions and local technological trajectories. Political tensions between stakeholders, negotiation and disorganisation often feature and are indeed the object of our analysis from the outset (see also Kallinkos, 2005).

We are particularly aware that our research problematic produced the framing that we adopt in this book. Hence we took account of a wide range of issues when considering how participating organisations were affected by the schemes, and we were cognizant of the mediating role of technology and 'the nature and rhythms of existing organisational relations' (Ribes et al. 2013, p. 11). There are many realignments which occur: Although empirical evidence of their impacts on organisations is scarce, drawing on the literature reviewed above we can surmise what Knights and Morgan (1997: 40) term the 'conditions of possibility' of the schemes as they unfold within the organisations we examine will be shaped by a number of different organisational dimensions:

> *Stakeholder interests* are implicated: those with an active political interest in the organisation's existing activities will exert influence as the power balance between them is renegotiated while the firm reconfigures its activities around government demands. Competing pressures will emerge between the government and the organisation; within supply chains, between the organisation and its customers and also its employees.
>
> *Information systems and infrastructures* are impacted in that they comprise the hardware and software which feeds government with securitised consumer data; as the regimes involve new information flows out of the organisation, existing information infrastructures will be re-purposed to accommodate government demands. Existing systems delineate the technological possibilities surround organisational responses. They will need to be adapted which will in turn impact the working patterns of people who use these systems.
>
> *Strategic adaptations* are made to these macro level changes as strategy makers respond to the macro-level phenomena of regulation and formulate an organisational response. They will renegotiate various

strategies to ensure that the firm continues to make money while absorbing a costly regulatory compliance burden. This renegotiation process takes time, compliance is a long term investment and the competing pressures are never extinguished.

Customer management processes are also involved: as more information is required of customers the customer relationship management process will be put under strain. While firms are keen to retain their customers they also have to incorporate the notion that every customer is a potential security threat into their customer relationship management processes.

Front-line workers: as front-line staff face the customers from whom security data are being collected and suspicion is being generated, it will be those staff who juggle the commercial priorities and regulatory burdens on a daily basis.

The conditions of possibility for the emergence of both AML/CTF and eBorders are thus numerous. A widespread concern for security in the wake of international terrorism, a belief that surveillance data gleaned from the general population will reveal the culprits, rising government debt, a neoliberal regulatory stance which simultaneously responsibilises organisations and strengthens the reach of the state in matters of public interest, and an approach to governance which relies on outsourcing to reduce costs, each feature in the contextual backdrop. There have been many – what business academics would call – macro-level commentaries on the significance of the private sector's involvement in national security, and movements in national security approaches more generally. Such commentaries have stemmed from the disciplines of political geography, international studies, political sociology and security studies. No study has, as yet, viewed these developments from the point of view of the private sector organisations concerned. At the beginning of this chapter we explained that government had diffused security issues into the private sector by requiring organisations to generate a securitised information flow to government about their customers. If we accept that security becomes depoliticised as it is diffused into everyday life, organisations, as intermediaries and generators of this securitised information flow, become important sites of potential re-politicisation in terms of what they are being required to do. As the public interest of security becomes inter-twined in the market driven logics of the private sector, what kinds of questions should be asked and what sort of perspectives should we employ? What happens within organisations when they are required to form part of the national security infrastructure, when their employees, customers, information systems and business models are seen as an extension of the security state?

The following chapters will address these issues as they arose in the empirical data and, critically, as organisational elements of each new governmental surveillance schemes demand co-operation and alignment. We incorporate literature which relates to each of these discrete business areas to explain the phenomena we observe and to feed into our analysis.

A note on method and data sources

We collected data using a mixed methods approach, featuring key informant interviews, two small scale industry surveys, and case studies in the retail financial services and retail travel sectors. In financial services, seven different industry bodies promoted the survey and we received 281 responses out of which 85 were complete. We also completed 28 interviews, 19 of which featured key informants from the sector (regulators, industry associations and MLROs and marketing executives). In depth access was secured to two organisations: a bank (company A) and a building society (company B). A total of nine case study interviews from these organisations were completed. Access was extremely difficult to achieve due to massive structural upheavals in the financial services sector following the banking crisis in 2009–2010 and the ongoing recession in the UK. The subject matter of our research was also sensitive for many organisations. In each financial services organisation we interviewed Money Laundering Reporting Officers, financial crime team staff and staff in high street branches or call centres. The interview guide probed the interviewees' understanding of AML/CTF, the impact of the regime on their organisation, which stakeholders were involved and how they were influenced, as well as operational adaptations that were made as a result of the regulations. Insight and opportunities for reflection were also provided by information gathered from corporate and media documents, attendance at industry events as well as in a stakeholder workshop the project organised for those who had participated as key informants.

Within the retail travel sector, responses to the survey were gleaned following a series of articles in the travel practitioner publication, Travel Weekly. Again we received 280 responses, of which 79 were entirely complete. We also completed 28 interviews, 13 of which featured key informants from the sector (typically regulators, industry associations and executives). With the help of the Association of British Travel Agents (ABTA), access was secured to three organisations: a large tour operator (company C), a large travel agency (company D) and a small travel agency (company E). A total of 15 case study interviews from these individual organisations were completed. The individuals interviewed at operational level represented a structured sample of managerial level marketing, IT and regulatory staff who had oversight of eBorders compliance, as well as front-line customer-facing staff either in high street branches or call centres. The interviews took place from early in the implementation phase of eBorders and for 12 months afterwards, enabling the developing situation to be understood. An interview guide was developed which probed the interviewees' understanding of eBorders, the impact of eBorders on carriers, the stakeholder groups and individuals which influenced or were influenced by the carriers, the extent and nature of interconnections between stakeholders and the context, processes and impacts of the eBorders programme. The interviews were transcribed and then coded and analysed using Nvivo software to enable patterns and themes to emerge from the data. The main bodies of coding were completed in team coding workshops, where we worked in small groups to code sections of data and then validate each other's coding practice. We completed a thematic analysis of the qualitative data (Boyatzis, 1998) but in places we also used some *a priori* coding derived from the theories and perspectives we adopt in the different chapters.

CHAPTER THREE

Shaping the regimes

Stakeholders and their interests

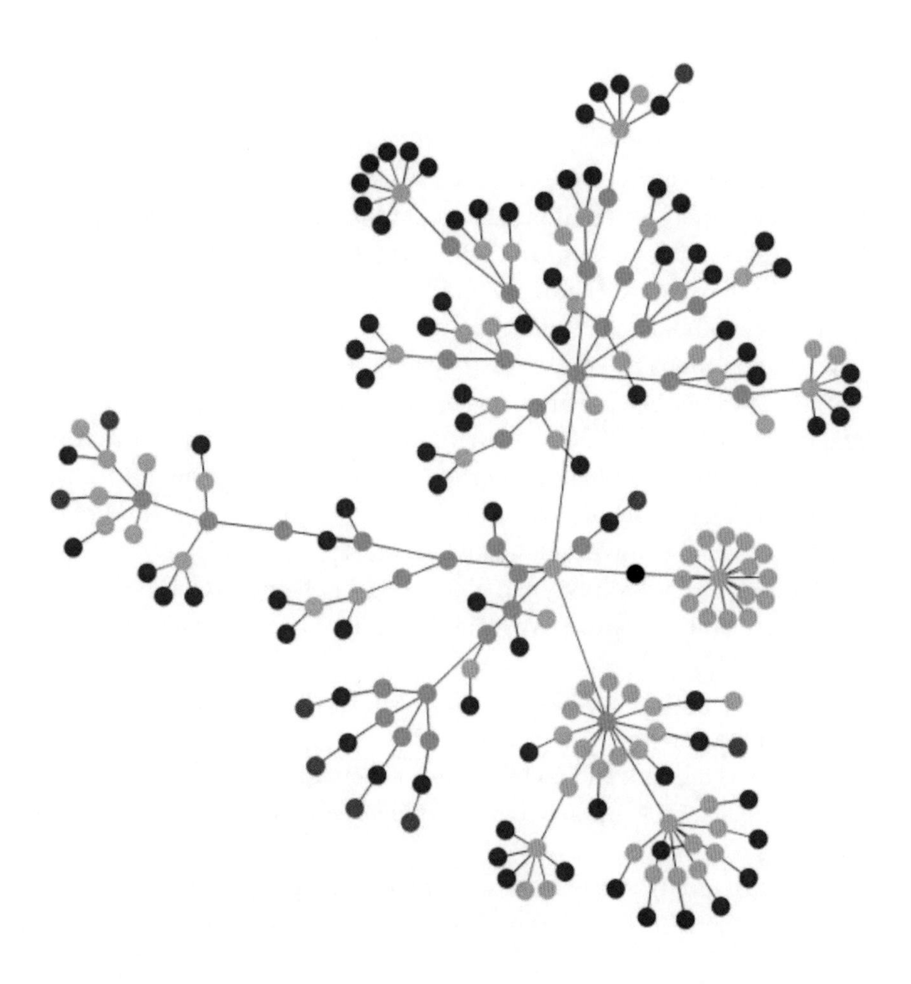

Introduction

In this chapter we identify and discuss the roles of the different stakeholders within each scheme. By considering stakeholder interests we lay the groundwork for an examination of how the information demands of eBorders and AML/CTF regimes introduced new tensions into existing organisational realities. It also enables an exposition of those who were influencing, and were influenced by, the regimes. We also examine the changing context in which the stakeholders operate and the interconnectedness between them. Therefore, this chapter uses a stakeholder-analytical lens (Freeman, 1984) which encourages the identification of all individuals or groups that influence, or are influenced by, an activity or collaboration. Consideration of the role, power and interconnectedness (Ackermann and Eden, 2011) of the stakeholders involved often identifies high degrees of influence between stakeholders as well as on the firm itself. This influence and interconnection will depend upon the environment or context in which the two regimes are developed and operated, and will both influence and be influenced by changes in that context (Pouloudi and Whitley, 1997).

We draw on data collected from multiple sources, via in-depth interviews, observations taken at industry events and seminars and document analysis. The interviews focused on key informants within regulatory, commercial and industry bodies allowing us to access the experiences of a diverse range of individuals and groups who can tell their own stories or 'epilogues' (Dibbern et al. 2008: 343). We worked around the stakeholder networks in a progressive and iterative sense, capturing their diverse views (Pouloudi and Whitley, 1997) and generating a rich understanding of their experiences of the schemes. We focus our analysis on two questions:

1. Who are the stakeholders involved in the two schemes studied and how are they interconnected?
2. How is the context in which these stakeholder networks operate been changing and what is the influence of those changes?

We give a brief introduction to stakeholder theory before explaining the stakeholder networks of eBorders and AML/CTF respectively.

Stakeholder theory

The notion that organisations have a range of stakeholders who have legitimate interests to which management need to respond is attributed to

Freeman (1984). According to Freeman, stakeholders are all groups and individuals that influence or are influenced by the organisation, resulting in many organisations having a wide range and disparate set of stakeholders. The ideas inherent in stakeholder theory, or 'stakeholder management' as it is sometimes called, have gained widespread acceptance and has been applied in many areas of business and management. We use stakeholder theory in our analysis to highlight how the requirements posed by eBorders and AML/CTF changed the nature of stakeholder relations within and around firms in the travel and financial services sectors respectively.

For all initiatives involving multiple actors, identifying key stakeholders is pivotal, a process which is shaped by issues of legitimacy and power (Jamal and Getz, 1995). Legitimacy relates to a particular stakeholder's right to be involved (Gray, 1985) and to whether they have the necessary skills and resources to do so (Jamal and Getz, 1995). The distribution of power across stakeholders, as in our two cases, tends to be uneven, with more powerful players threatening the interests of the less powerful (Archer, 1995), who may struggle to get their interests considered (Tosun, 2000). Stakeholder analysis must therefore take into account the conflicting or negative aspects of stakeholder relations (Friedman and Miles, 2002). Whilst stakeholder theory was formulated at the level of the entire organisation and is often applied at that level, scholars of information systems (IS) have tended to apply it at the level of the system being considered. There is certainly an information systems element to both of our cases. There is widespread recognition that system development, implementation and use is more likely to be successful if a wide range of stakeholders, and not simply the system developers and users, are included in system development and implementation (see for example, Benjamin and Levinson, 1993; Kumar et al., 1998; Ward et al., 2005). Similarly, examples abound of system developments that failed to be completed or that failed to deliver the expected benefits because of lack of stakeholder engagement (e.g. Pan, 2005; Mantzana et al., 2006).

Stakeholder interconnectedness

The breadth of the stakeholder concept, allied to pressures such as globalisation and the ubiquity and speed of communications (Fassin, 2009), results in organisations being able to influence, or be influenced by, a broad range of stakeholders. The original stakeholder model presented by Freeman (1984) shows the focal organisation as a hub, with the stakeholders viewed as spokes. Intuitive as this view may be, it fails to acknowledge that stakeholders also influence each other (Fassin, 2009) and that a combination of stakeholders can exert more influence on a focal firm than each separately (Mitchell et al., 1997; Friedman and Miles, 2002). This suggests that it is important to explore the interconnections between stakeholders.

In the particular case of IT based activities, Pouloudi and Whitely (1997: 3) note that whilst a small scale IS may involve a limited set of stakeholders, organisation-wide and strategic IS will require involvement from a wider set of stakeholders from across the organisation and when considering inter organisational information system (IOS) 'the number of stakeholders involved in systems development and use is far greater than that of most traditional [intra-]organisational systems'.

Importance of context

As part of the ebbs and flows in the process of organising, specific stakeholders can become more or less salient over time as a result of, for instance, new coalitions between stakeholders (Mitchell et al., 1997) or changes in the environment (Phillips, 2003). Therefore, stakeholder analysis may benefit from the adoption of a specifically processual lens and needs to consider how stakeholder relations change over time, as well as how and why such changes occur (Friedman and Miles, 2002; Pouloudi and Whitely, 1997). The widespread use of stakeholder theory has led to detractors, such as Trevino and Weaver (1999) who assert that stakeholder theory is not a theory, but a research tradition. In order to address such criticisms, Donaldson and Preston (1995) seek to explicate three perspectives within stakeholder theory: descriptive, instrumental and normative. The descriptive perspective seeks to identify the stakeholders involved in an organisation and the interactions between them. It is this aspect of stakeholder theory that often attracts criticism, since it provides limited insight and explanatory power. The instrumental perspective links stakeholder theory with improved organisational performance. Finally, the normative perspective links stakeholder theory with business ethics and corporate social responsibility, recognising that incorporating stakeholder interests is a worthwhile end in itself. Bailur (2006: 65) notes that the normative perspective may be particularly relevant to IT developments where firms have 'increased responsibilities' (quoted from Reed, 2002). For the firms we investigated, such responsibilities now clearly involve the transfer of customer data to government. Other criticisms include that there is sometimes an 'excessive breadth in the identification of stakeholders' (Donaldson and Preston, 1995: 86), noting that there is a distinction between those that have a stake in the firm (i.e. that they stand to benefit from the success of the firm) and those that influence the firm or are influenced by it, but who have no direct stake. It is inclusion of all possible 'influencers' (p. 86) that lead to the excessive breadth in the identification of stakeholders noted and the seemingly impossible managerial task of managing such a breadth of stakeholders. In our analysis we adopt Donaldson and Preston's (1995) view, suggesting that the successful implementation of both AML/CTF and eBorders schemes is contingent on the economic success of these enterprises, and identifying stakeholders accordingly.

Identification and interconnectedness of stakeholders

Consistent with stakeholder theory, Figures 3.1 and 3.2 show that participation in the schemes results in the focal firm influencing or being influenced by a number of stakeholder groups. Furthermore, these stakeholders have inter-relationships between them, consistent with the notions of 'stakeholder's stakeholders' and stakeholder networks suggested by Pouloudi and Whitley (1997). The introduction of these new information infrastructures and the new organisational practices they supported began to reconfigure existing stakeholder relationships, placing them under some strain. In this section we examine the nature of those stakeholder relationships. There are some notable similarities in that the government is a key, powerful stakeholder, having passed the legislation which mandates the existence of these two schemes as well as the terms of compliance. However there are also some key differences.

Travel sector

The eBorders stakeholders and their interests are described in Table 3.1.

Stakeholder	Stake/Interest
Airline – legacy	Airlines are responsible for transmitting advanced passenger information (APIS) to UKBA before travel. Information must be gathered at booking, separately after booking or at check-in; adopting technical standards imposed by UKBA. Some carriers can use their loyalty scheme to record APIS data of customers.
Airline – charter/budget	Like legacy carriers, charter and budget airlines must transmit APIS data to UKBA. Charter flight bookings are often made in groups without data on individuals being provided.
Airport operators	Those providing check-in services for airlines must ensure APIS data is collected and transmitted on behalf of the relevant airline.
Tour operators	Not ultimately responsible for passing APIS data to UKBA, but operators worry about reputational damage if customers cannot travel. Where possible, they collect data at booking, with some providing websites for customers to provide APIS data online.
Travel agents	Not ultimately responsible for passing data to UKBA, but agents also worry about reputational damage if customers cannot travel. Originally encouraged customers to provide APIS data to tour operators or airlines, but increasingly concerned about losing the customer relationship.
Industry bodies e.g. ABTA	Represent the views of industry groups, and have made representations about the legality of eBorders in Europe. Concerned about technical aspects of data format and transfer, to ensure the use of existing industry standards and to minimise additional costs.
IT suppliers (Trusted Borders)	UKBA formed a consortium of IT hardware and software suppliers to develop the data transfer, storage and analysis system. In July 2010, the programme's main IT contractor was sacked (Millward, 2009; Ford, 2010), but was replaced in April 2011.
Global Distribution System Operators	Enable their users (airlines, travel agents and tour operators) to input APIS data quickly and efficiently. Successfully lobbied by ABTA to add lines of script which dealt with APIS input.

UK Border Agency	Agency is responsible for visas, managing immigration and border security, and for developing/operating eBorders. Responsible for examining the supplied data to identify and investigate suspicious activity and individuals, including monitoring against watch lists.
UK Government	eBorders was introduced by the Labour government in 2003, originally as a means for immigration control. After the London 2005 bombings, security and counter-terrorism became additional rationales. The coalition government formed after the 2010 election continues to support the programme.
European Commission	Several stakeholder groups, including the airlines and industry groups made representations to the EC about the programme's legality under EU laws on free movement. EU citizens were judged to have the right to withhold data when travelling within Europe.
Customers	Customers must provide their APIS data prior to travel. Awareness and understanding of the scheme are low, with many unaware that information is needed or that it is being recorded, analysed and stored by UKBA. Even fewer customers are aware of their right to opt-out of the scheme for intra-EU flights.
Source: Dibb, S., Ball, K., Canhoto, A., Daniel, E., Meadows, M. and Spiller, K. (2014). Taking Responsibility for Border Security: Commercial Interests in the face of e-Borders. *Tourism Management*, 42(1), 50–61.	

Table 3.1. Stakeholders in the eBorders programme.

Table 3.1 describes the key stakeholders in the eBorders programme and their interests. We also represent this graphically in Figure 3.1. The key stakeholders in the eBorders programme represent the travel supply chain, multiple layers of governmental stakeholders in the UK and the EU and individual customers. It is possible to extend beyond the boundaries of the EU as the US government's USVISIT programme was part of the inspiration for eBorders. In order to create a securitised information flow from the travel sector, the government, in this case the UK Borders Agency (BA) has established requirements for the type of data that must be collected and transmitted and when this must occur. The UKBA established a consortium of IT and hardware suppliers to build the data transfer, storage and analysis system. These suppliers have a relationship both with the UKBA and the focal firms that must develop interfaces with their systems. This interfacing between systems has required considerable discussion and in some cases investment from the focal firms and has, in part, been undertaken on behalf of travel firms by pan-industry bodies such as the Association of British Travel Agents (ABTA). As noted above, the government has recently sacked the major IT contractor in the consortium of suppliers, which as can be appreciated from the degree of interconnectedness shown in Figure 3.1, has resulted in uncertainty, delays and increased expense across the majority of stakeholder groups.

There is a high degree of interconnectedness between industry partners because of the long retail travel supply chain. Airlines use agents and tour operators to sell seats on their aircraft as well as selling them directly to customers. Whilst many travel operators sell directly to customers, an approach that grew rapidly in the e-commerce boom of the early 2000s, there is still considerable use of agents and partners in the travel sector. The best known of these is travel agents, which sell both complete holidays and individual travel components on behalf of tour operators, airlines and other travel providers. Even if a travel firm does not use a travel agent, it may have relationships with other partners. For example, frequently a tour operator running a charter airline will work with 'sharers' which are other tour operators that have block booked seats on their aircraft. Similarly, they may not undertake operations at the airport themselves, rather they will use a third party ground handling company. Therefore, the focal airline has a relationship with these agents and partners, and similarly they have a relationship with the focal firm. These agents and partners are often used in order to provide an interface with the customer, hence these organisations also have a reciprocal relationship with the customer or traveller.

Much of the work in responding to eBorders directives from government was undertaken on behalf of individual travel providers by cross-industry working groups. These groups interacted both with the consortium of suppliers contracted by the government and also with the government themselves, in order to provide feedback on design and implementation of the scheme.

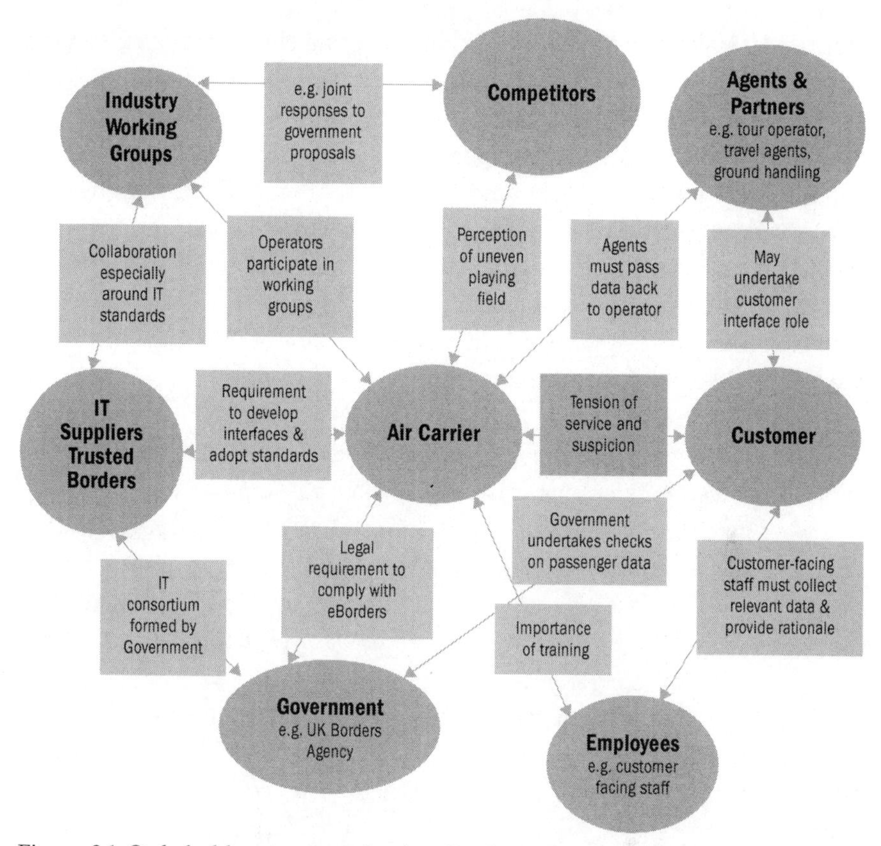

Figure 3.1. Stakeholder structure for the eBorders scheme.

It is the travel operator's ultimate responsibility to collect data from customers and transmit this to UKBA's data centre. They may carry this out directly, for example by using web portals, or if they are using a travel agent or ground handling firm, this information may be passed via this agent or partner. However the data is collected, it is most likely to be defined by the operator's processes and handled by their IT systems, hence the focal firm has a stakeholder relationship with the customer. Whilst the focal organisation's processes and systems will affect the customer, it is the customer-facing staff, such as staff in call centres, that will interact directly with customers and will be required to ask for the data required and explain why this information is needed and how it will be used. Hence we reflect in

Figure 3.1 a stakeholder relationship between front-line staff and customers. Equally, these staff have a relationship with the focal organisation, which is their employer, based on many things including adherence to policy and processes and performance management. Due to its recent introduction, key informants interviewed in the tour operators identified training as a key part of how staff can effectively carry out their roles, both for their own satisfaction and for the benefit of the firm and other stakeholders. However, as described by an informant from a tour operator that was required to help customers enter their travel information into the firm's online form, the training has in cases been very limited:

> There wasn't really any training. It would have been an e-mail from the boss at the time saying that an e-mail had come from [location of head office] saying we are now to take these calls...we all had a little practice on the test website and that was it. Nothing to do with rules and regulations, but just what to put into the website. That was it as far as the training has gone.
> (T13, Call Centre Team Leader, Tour Operator)

As described previously, the focal firm is required to pass information to the UKBA, who undertakes checks on the data and acts on suspicious activities. Hence the government agency and customers also have a stakeholder relationship.

Finally, Figure 3.1 shows that competitors should also be considered stakeholders in the eBorders programme, since they can both influence and be influenced by the focal firm. As mentioned previously, in order to influence the development of eBorders, particularly with regard to the development of IOS issues such as data standards and the development of interfaces, competing travel operators have collaborated in pan-industry trade bodies. As also mentioned above, competitors may also act as partners, for example, when sharing aircraft or other forms of transport of infrastructure, such as airports, ports or terminals. Despite the levels of collaboration and partnership, interviewees also described significant disparity between the size of different travel operators and in their legacy systems. Similar to the financial services sector this has resulted in the perception of significant power imbalances between competitors and the perception that some larger operators may have a greater say in how the eBorders systems are developed.

The changing context of eBorders

Interestingly, whilst the media reports identified the control of immigration as the initial context for the eBorders programme (Woolas, 2009), staff in the travel firms interviewed viewed the scheme as being addressed solely at increasing national security. Most interviewees were sympathetic to increasing security, since it is seen as both a common good and also of benefit to the travel sector, which is often impacted by security threats. However, there was also recognition that their customers were highly sensitive to being asked for personal information and wished to know why that information was being requested and how it would be used. For example, staff in a tour operator described how they had developed an online form to capture the information needed from customers for eBorders. However, they had also had to set aside one section of their call centre to answer questions from customers about the online form and the data required. Whilst some of these questions were purely operational, such as lost booking reference numbers, a good number of calls involved the customer questioning why this information was required. Rather than raise issues of national security with customers, the call centre staff interviewed as part of the study described how they sought to downplay the matter in order to encourage customers to comply, describing it as 'just a government thing' and that all companies now asked for this information and that providing the information online 'would save time at the airport' [Interviewee 10, Tour Operator]. This approach to downplaying the rationale for collecting the information appeared to lead some staff being inaccurate with what they told customers, for example one key informant said if they asked, customers would be told 'generally it is because the government of the country you are going to needs this information', rather than that is was the UK government that required the information.

Size was also an issue, as one interviewee explained:

> When you're a small independent ski tour operator that buys ten seats from [name of airline] a week, and that's the entire size of your operation, your business is run out of your back bedroom, the IT changes to make all this happen are just too high, the costs just don't stack up.
>
> (T4, Business Development Executive, Airline)

Smaller travel agencies also spoke of their frustration about the decentralised elements of eBorders and the responsibility that was placed on them for collecting and transferring data. Some complained that it was easier for legacy carriers to adapt their systems to cater for eBorders, while others saw the required investments in systems, staff and training as a serious problem. Smaller firms were particularly concerned about their margins.

Financial services

Table 3.2 shows the range of stakeholders in the AML/CTF regime. This is also shown graphically in Figure 3.2.

Stakeholder	Stake/Interest
Financial services institutions	Financial institutions in the UK face a regulatory requirement to observe due diligence with respect to their customers' account activity, and they face unlimited penalties if they fail to do so. Financial institutions bear the responsibility for customer vigilance, but have no say in public policy or the development of regulations.
Employees	Employees of financial services organisations must not knowingly assist a money launderer or terrorism financer, they must report suspicions of money laundering and terrorism financing, and they must not, under any circumstances, let the customer know that a report has been made about him or her. Failure to comply can result in prosecution (between 5 and 14 years imprisonment) and unlimited fines. Staff should only discuss their suspicions with nominated members of staff – namely, the firm's Money Laundering Reporting Officer (MLRO), and members of the AML/CTF team. The MLRO is the person in charge of money laundering and terrorism financing detection at each institution. He or she is criminally responsible for the institution's AML/CTF compliance.
Independent financial advisors	They must confirm the identity of their customers, and must not knowingly assist a money launderer or terrorism financer.
Industry bodies e.g. BBA, JMLSG	Represent the views of industry groups, and have made representations to the regulatory body about aspects of data collection and analysis. They also develop guidance notes for financial institutions.

IT suppliers	The regulator does not mandate the use of AML/CTF software, but often implies that financial institutions can not rely solely on staff to manually check customers' identity against blacklists and monitor transactions for suspicious activity. The specific IT solutions adopted by each financial institution vary widely, and may include vendor solutions, employment of in-house data miners to develop ad-hoc queries or the adoption of models developed by financial intelligence units.
Law enforcement agencies	When financial institutions identify suspicious activity, they submit a Suspicious Activity Report (SAR) to the National Crime Agency (NCA). NCA then decides which law enforcement agencies (LEAs) should be informed (e.g., local police forces, customs, UK Borders Agency, etc) for further action. In addition to investigating SARs, LEAs can raise court production orders, demanding that financial institutions provide witness statements and evidence regarding particular customers under investigation for AML/CTF offences.
UK Government	At the time of the study, the Financial Services Agency (FSA) regulates the financial services industry in the UK. Since 2002, it has formal powers to supervise and enforce compliance with laws and regulations relating to AML/CTF. The legal requirements were further strengthened with the introduction of the Proceeds of Crime Act (PoCA) in 2002 which extended the definition of money laundering and created a new imprisonable offence of failing to disclose suspicious transactions in respect of all crime. The FSA puts forward two instruments for AML control – know your customer (KYC) and transaction monitoring – and conducts regular audits of the industry. The FSA can impose unlimited fines on institutions that fail to comply with AML/CTF laws and regulations.
European Commission	The European Commission's AML directive provides the common legal ground for all member states. The directive is transposed into national law and regularly updated (e.g., revising the definition of money laundering). The directive adopts a minimalist approach in order not to collide with the exclusive competence of member states and to accommodate national interests such as secrecy laws. Consequently, the transposition into national law has resulted in uneven applications among member states.

Customers	Customers must provide proof of their identity, including their address. Under the terms and conditions of acquiring a financial product – e.g., opening a current account – they must also agree to their financial transactions with that institution being monitored. Customers are not informed when they are being investigated for suspicious behaviour, or when a report (i.e., a SAR) has been made about them.
Transnational bodies e.g., FATF, Basel Committee, Egmont group	Supra-governmental bodies were established to provide governance on a global scale. They develop AML/CTF measures and recommendations for governments, financial intelligence units and law enforcement agencies across the worlds; or for professionals in particular industries (e.g., the International Federation of Accountants). Some also produce 'blacklists' of governments or institutions deemed not be doing enough in terms of preventing and detecting AML/CTF.

Table 3.2. Stakeholders in AML/CTF.

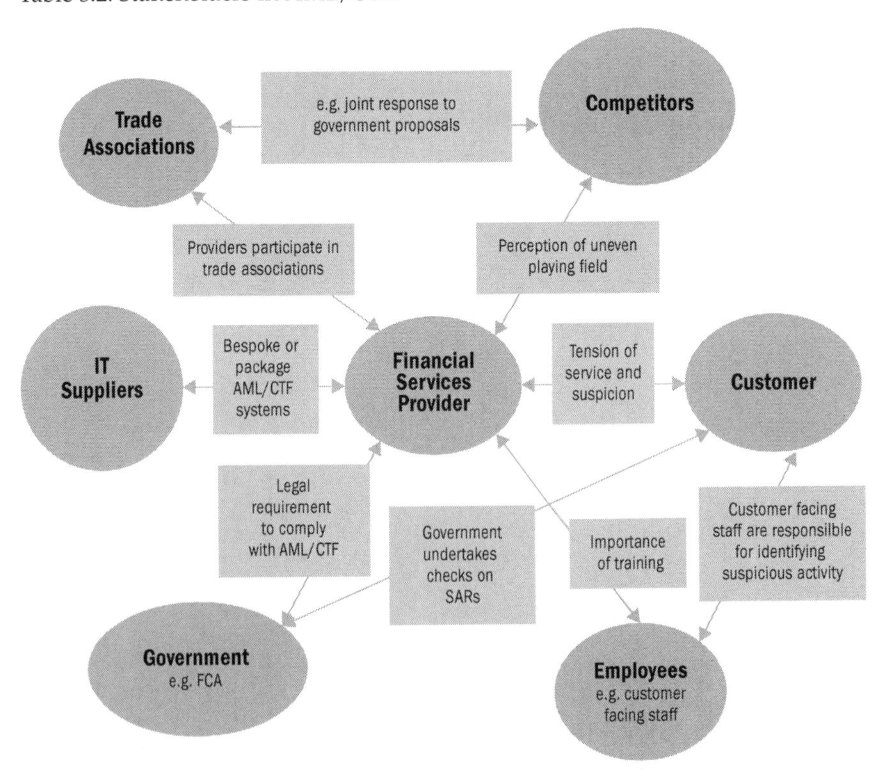

Figure 3.2. Stakeholder structure for the AML/CTF scheme.

As described in chapter two, a key difference between AML/CTF and eBorders is that financial firms are required to identify suspicious transactions themselves rather than simply pass data on. Focal firms in this sector also do not use agents and partners to sell their products. The information flow in financial services – which takes the form of Suspicious Activity Reports – is therefore less voluminous and more the result of firms' own risk analyses of customer behaviour. The National Crime Agency (NCA) investigates suspicious activity reports (SAR) and decides if and how to pursue customers directly. Hence the government and customers have a stakeholder relationship. Whilst AML/CTF does affect customers, front-line staff, such as branch and call centre staff, also have a stakeholder relationship with customers. Key informants noted that training was a key means of enabling customer-facing staff to carry out their responsibilities under AML and CTF regulations. In contrast to the travel sector, informants in the financial services sector described how training relating to AML and CTF is both extensive and is renewed and updated on a regular basis.

Informants described how they had either developed information systems to undertake AML/CTF activities in house, or had bought commercial systems from the different suppliers in the market. Hence, unlike the travel sector, where the government created a consortium of IT and hardware suppliers that travel operators were required to work with, the financial services sector have been able to choose suppliers themselves or develop their own systems. Whilst both of these approaches have caused the firms to incur considerable costs, financial service firms were able to select systems that were best suited to their needs, for example systems that were compatible with their legacy systems and operations.

The financial services sector has a number of trade bodies that represent them to government and in other fora (e.g. the British Bankers Association, BBA). Since the financial services firms have been allowed to determine their own approach to system development, the trade bodies have not been required to work as closely with IT suppliers. In fact, there isn't even an explicit requirement for banks to use IT, although it's implied in the FSA AML policy. As with the travel sector, trade bodies consist of competitors seeking to collaborate to develop common approaches to non-competitive issues. However, key informants in the financial services sector noted that there were significant power imbalances between competitors resulting from different organisational size and hence access to resources and expertise in relation to their AML/CTF practices.

The changing context of AML/CTF

Informants from the financial services sector stated most customers were not aware of AML or CTF and hence the national security issues related to these. However they also stated that their customers recognised that activities such as proving identity when opening an account and making significant transactions was in their own personal interests since it reduced the possibility of theft or fraud from their account. A typical response from an informant in this sector described how the wider context of consumer concerns about privacy was helpful:

> Customers are concerned about things like fraud and identity theft and this actually helps us meet our obligations under AML. Customers don't mind providing proof of identity or if someone from the bank calls them back to check on an unusual transaction. Often the customer is reassured by this – not annoyed.
>
> (F13, Marketing Data Manager, Retail Bank)

An interesting change in the financial service context was that interviewees noted that the global financial crisis had put the operation of financial services providers under even greater scrutiny. They therefore described how AML/CTF issues were given even higher priority than before the crisis, since they did not want to be seen to make further mistakes. This was expressed particularly strongly by one of the retail banks in the UK that has been taken into part public ownership. This demonstrates how seemingly independent elements of external context, such as national security and the global financial crisis, can combine to change the context, in which stakeholders operate.

Conclusion: Reflections on stakeholders

Using the theoretical frameworks of stakeholder analysis we have begun to paint a picture of how the eBorders and AML/CTF began to intersect with organisational processes and resources. Stakeholder analysis has allowed us to identify how existing stakeholder relationships emerged in the regimes and to explore their changing context.

Figures 3.1 and 3.2 show the key stakeholders involved in each of the schemes of interest and the high degree of interconnectedness between them. Whilst previous studies of IOS development have discussed interconnections between stakeholders (for example, Pouloudi and Whitley (1997) draw on notions of social and industrial networks when they discuss the interlinkages between stakeholders), published studies tend to presents lists of stakeholders, rather than depict their interconnectedness. Both sectors studied show high levels of interconnectedness, suggesting that it is not particular to one sector. Rather, consistent with studies of the healthcare sector (e.g. Boonstra et al., 2008), our findings suggest that information systems, particularly those which span organisational boundaries and which are developed to address government priorities, are likely to have multiple interactions between stakeholders. Customers, for example, will be both stakeholders of the focal firm and also citizens of the government. Interestingly, in eBorders, the government had commissioned a consortium of IT suppliers to develop the government's information system, which led to a greater degree of interconnection between stakeholders than in the more established AML/CTF scheme. We would suggest that if government continues to dictate, not only the actions of private sector organisations but also how their information systems should operate, then organisations will experience a greater degree of stakeholder interconnectivity which might begin to impact competitive strategy. The implication of this greater degree of stakeholder interconnection is that the focal organisations cannot simply manage the relationships with their stakeholders, but must be alert to the relationships between stakeholders, which may influence their activities and operations. The more interconnected stakeholder relationships become, the more there is potential for greater complexities. Any change in these interconnected relationships will then have more complex consequences and as we discuss in the following chapter, the greater opportunity for tensions to arise.

Our findings also show that the changing context has had an impact on the development, attitudes towards and use of the schemes studied. This includes influence from events outside of the direct sphere of interest, such as the global financial crisis and rising awareness of surveillance and the value of privacy. These widespread and diffuse influences are often ascribed to the influence of the media, but we would assert that they are much more widespread and include: word of mouth between consumers, education and even the actions of national governments themselves (Huysmans, 2006). We therefore support the idea that the wider context, including the changing nature of that context, should be explicitly recognised when undertaking stakeholder analysis and considering the influence of stakeholders on each other. We believe that this recognition of the wider context is particularly pertinent to IOS development in fields such as national security and personal privacy, since these domains draw on the deep-rooted personal views and beliefs of individuals and the society in which they live, which can be influenced by a wide range of contextual factors.

In the next chapter we begin to examine stakeholder views on the regimes and their information demands. Stakeholders in both sectors described the disruptions and tensions which surrounded the production of the securitised information flows required.

CHAPTER FOUR

Secure information flows?

Tensions, disruptions and realignments in information infrastructures

In order to produce a securitised information flow to serve the regimes of eBorders and AML/CTF, large scale information infrastructures were developed. These infrastructures, which we term 'inter organisational systems' (IOS) spanned organisational boundaries and connected them with government. They allowed the collection, analysis and transfer of the large quantities of data that the regimes required. The introduction of these IOSs introduced new tensions into existing stakeholder relationships because of the adjustments in infrastructure the IOS represents.

Our data reveal that the new tensions introduced by the schemes run right through the organisation. Predictably, the tensions which emerged did so in different forms and it was interesting to observe their ongoing nature. In travel, tensions associated with the introduction of eBorders pervaded competitor and supply chain stakeholder relationships as well as within the firms. In financial services, the mature AML/CTF scheme exhibited fewer tensions at macro strategic levels but more at operational level and those tensions remained within the firm. The tensions we observed signified that the securitised information flows associated with each regime were causing disruptions and realignments in organisational resource use. Compromising the interests of stakeholders, it also took on a political character. In this chapter, we adopt an analytical lens from the field of information infrastructures (IIs). We offer a brief overview of the II literature before exploring the data.

Information infrastructures

Information infrastructures (IIs) describe the technical and human elements of information based systems (e.g. Gal et al., 2008). Whilst information technology (IT) is not a pre-requisite for IIs, in practice, many of today's IIs have developed around IT systems (Henfridsson and Bygstad, 2013). IIs are deeply embedded in organising processes and often only become visible when they contribute to contested outcomes or breakdowns. They both shape and are shaped by the conventions of the organising processes in which they are embedded and often involve interconnected IIs, inheriting their strengths and limitations (Star and Ruhleder, 1996). IIs require connections and reciprocal actions between a range of stakeholders (Ribes et al., 2013). They create and support specific configurations and actions, while disabling others (Star and Ruhleder, 1996) and can change the process of decision-making (Ciborra, 2000).

Development and implementation of IIs are intended to produce a required outcome. Often this is related to an increase in control over an activity or process of interest (Ciborra, 2000). For example, in the schemes studied the IIs are intended to increase control over the movement of money within the UK economy and of individuals across the UK border, with the assumption that this will improve national security (Salter, 2004; Gill and

Taylor, 2003). Use of IIs has also been associated with unintended consequences such as work-arounds (Boudreau and Robey, 2005), resistance (Rivard and Lapointe, 2012) and reduced control due to the paradox of control inherent in complex II (Hepsø et al., 2009). As in the iterative relationship between the structure and use of IIs and the tensions that arise, an interaction can be expected between intended and unintended consequences. For example, if staff do not support the intended outcome of the II they are more likely to resist it (Polites and Karahanna, 2013) and such a movement often creates more problems within the organisation.

Tensions arising from information infrastructures

Tensions occur in the development and operation of IIs as a result of the different interests, stakes or participants (Ure et al., 2009). These interests come into particularly sharp relief when forms of centralised control are attempted within the II (Hepsø et al., 2009). Studies have identified three types of tensions surrounding IIs. The first category of tensions relates to technical issues (Henfridsson and Bygstad, 2013). The agreement of technical standards, in particular, can cause problems if the standards differ from existing practices (Braa et al., 2007, Vikkelsø, 2007). The second category relates to the costs of participating in the II (Bowker and Star, 1999). Costs may be unevenly distributed across stakeholders leading to feelings of inequality between them. Cost-based tensions are often accentuated in the on-going operation of IIs, even if not present in the initial development phase as 'funds for developing flashy new systems are relatively easier to come by, but on-going money for maintaining operational systems (the workhorses of domain science) is in vanishingly short supply' (Edwards et al., 2009: 367). The third category of tensions concerns the context in which the II is deployed. Technical actions take place within organisational and social contexts, with their own practices and culture, often with deep heterogeneity at the local level (Ure et al., 2009). The II literature suggests that the three types of tension identified are inter-related or embedded (Ribes and Finholt, 2009). For example, the requirement to adopt certain operational standards may generate technical tensions, give rise to costly changes and may favour the working practices of certain groups over others (Bowker and Starr, 1999). In a theoretical discussion Kallinikos (2005) makes a general, but very important point that it is in the nature of an information system, particularly one which spans organisational or geographical boundaries to operate under conditions of functioning simplification (Luhmann, 1993). They operate to unify what he terms the 'information habitat' of organisations, closing down local variation and establishing wider standards. Because of their widespread interconnectedness, any tensions and disruptions are likely to be far reaching with unforeseen knock-on effects. In short, the stakes are higher: with the greater investment in centralised control and interconnectivity comes the potential for widespread chaos should things go wrong. The extraction of customer information from firms for security purposes required a decoupling of that information from its operational and organisational context. Yet that decoupling was hard won and produced numerous consequences for stakeholders.

Securitised information infrastructures

The changes in practice which were required so that firms could adapt to the eBorders and AML/CTF requirements were outlined in chapter two. We describe the specific information systems which made up the security information infrastructures in firms, to support these changes in practice, in the following pages.

Travel sector

In the travel sector, compliance with eBorders required airlines to gather and transmit passport information from customers between 24 hours and 30 minutes of departure. Because airlines were not the sole vendors of seats on their aircraft, and different types of airlines used different types of information system, complicated chains of systems were linked down the travel industry supply chain. New systems were as follows:

- Airlines, travel agents and tour operators developed 'self service' web and smart phone portals (supported by customer service call centres) so that their customers could enter their passport information directly.

- Airlines developed 'self service check-in kiosks' which gathered passport information in the airport and transferred it to the UK Border Agency.
- Ground handling systems in airports (which were often rented by airlines other than legacy carriers and so had independent systems) incorporated new functionality to transfer passport information gathered at the check-in desk.
- Operators of the airline industry's Global Distribution Systems incorporated new fields where passport information could be input and transferred directly to the UK Border Agency.
- New connectivity standards to transfer passport data were developed between travel agents, tour operators, airlines and the UK Border Agency.
- The Trusted Borders consortium built the government eBorders data warehouse infrastructure as well as the systems which supported border staff in their identification of suspicious travellers.

Financial services

In the financial services sector, compliance with AML/CTF regulations relied on identifying and reporting suspicious activity when customers opened an account and subsequently when they used that account. As such, the firms we investigated had developed and interlinked the following information systems to support their AML/CTF practices.

New internal systems included:

- Customer Management Systems which enabled staff to identify the identity documents required from the customer and to report that identity documents had been gathered.
- Institution-wide fraud reporting systems so that staff could communicate suspicions to the internal financial crime team.
- Transaction Monitoring Systems which monitored all transactions in real time and screened all transactions on a daily basis for suspicious behaviours (identified through profiling and rules-based targeting). Transaction Monitoring Systems also searched for names which appeared on government watch lists which were disseminated to banks on a weekly basis.

Interlinked external systems included:

- Credit scoring systems used to verify customer identity and creditworthiness.
- The National Crime Agency's Suspicious Activity Reporting portal, to which all financial institutions had secure access in order to submit a SAR.

The differences between the two schemes meant that as well as introducing similar stakeholder tensions, there were some very different tensions as well. In both sectors, an overarching source of tension was the conflict of interest inherent in balancing regulatory and commercial obligations. Participants repeatedly referred to the nature of the regulation itself as inherently problematic and counter to commercial imperatives. Consistent with the consideration of tensions in extant II studies, it was found that tensions were, in some cases, embedded or interrelated and existed across multiple scales of action (Ribes and Finholt, 2009). For example, in the travel sector, the focal firms have been required to use a data standard that is more expensive than the one previously used. This has resulted in tensions both from the technical requirement to change systems and the additional costs incurred. In a few cases, the interviewees mentioned that once the appropriate systems, processes and training had been put in place, there may be benefits that could be realised from compliance. However, such views were in a minority. In addition to the overarching tension between commercial and regulatory imperatives, our analysis identified five specific areas of tension that are discussed in turn below. Each of these tensions is further considered according to its apparent underlying cause, as suggested by the II theoretical lens, namely *technical, cost, and context.* A summary of the analysis is provided in Table 4.1.

Tensions between organisation and government/regulator

The travel sector interviewees revealed that there were significant technical and contextual tensions which arose between government and the sector. Although some tensions arose because the scheme is still being implemented, the overall government approach to the implementation of eBorders also prompted difficulties. Interviewees in the travel sector felt that the technology suppliers had not listened to how the industry worked together at a technical level, preferring instead to listen to system suppliers:

> Because we do commercial business together our systems have to talk to each other … And they failed to recognise that aspect. They were having conversations with system suppliers rather than carriers.
>
> (T15, Senior Executive, Trade Body)

One issue that had caused particular tension was the choice of a non-standard data transfer protocol, SITA type B, which is more expensive than the more widely used UNEDIFACT protocol. SITA was one of the companies involved in the Trusted Borders consortium and appeared to have influenced the choice of the protocol. As one airline interviewee explained, this supplier was also able to promote their own check-in systems to airlines and airports around the world by saying that they were eBorders compliant:

> So SITA have been out selling to all of the airports saying you need our new all-singing, all-dancing check-in systems that are compliant with UK APIS.
>
> (T4, Business Development Executive, Airline)

In addition, the centralised and relatively high cost technology solution imposed on the travel sector had resulted in significant extra costs, as described by an airline employee:

> It turns it almost into a tax. Some of the Caribbean airports … they're trying to charge us $1.80 a passenger. You then multiply that across an operation or a flight and suddenly … it's now costing me, I don't know, $800 more than it was yesterday. So that either goes on the ticket … or we absorb the cost which I don't particularly want to.
>
> (T4, Business Development Executive, Airline)

The challenge from the EU and the requirement to allow EU citizens to opt out of the eBorders scheme has created additional tensions for the focal firms. Systems were designed to gather passport data on a mandatory basis, rather than to include an opt-out. If the EU were to insist on the opt-out then huge development costs would be incurred. As a result, some airlines have merely 'switched off' eBorders within the majority of European countries. Others continue to insist on mandatory data collection.

Furthermore, the UKBA has not mandated training about eBorders, which has resulted in piecemeal approaches to staff training.

This lack of systematic training in the travel sector appears to be compounded by a lack of widespread public awareness of the eBorders. The UKBA has adopted a low key approach to the launch of the scheme; for example, there has been no public information campaign. This has resulted in staff in the travel sector being required to both educate their customers and develop and implement the necessary systems and processes. As one interviewee described:

> One or two of them who maybe travelled last year know they are going to be asked for passport information. Some of them will bring it in, but not that many have cottoned on to it yet.
>
> (T19, Retail Manager, Travel Agent)

By contrast, AML/CTF is a longstanding scheme with the compulsory responsibility designated locally to participating firms. Despite the longevity of the scheme, tensions are prevalent in relation to some technical and contextual matters. An interviewee from the financial services regulator described how there were varying levels in the quality of compliance, which required them to implement remedial action:

> The banks tend to be very good because they spend a lot of money on training and have a huge setup which is pretty good, plus they're automated because they're dealing with so many transactions on a day. While the poorer standard of reports tend to come from the one-man band who doesn't understand fully his obligations under the legislation.
>
> (F20, Public Relations Manager, Government Agency)

For the 'one man band' cost, in terms of systems development and staff time, was a significant factor affecting compliance. Larger institutions did not experience such issues because the initial systems outlay had already been absorbed and they benefitted from economies of scale. However, while some interviewees recognised that their firms embraced social responsibility and actively contributed to addressing security challenges, others were less positive. As this interviewee explained:

> Ultimately banks are not interested in the good citizen role. There are no obvious benefits to the organisation of applying AML measures.
>
> (F7, MLRO, Retail Bank)

Tensions between organisations and their customers

In the travel sector, tensions arising between customers and firms related to technical and system implementation difficulties. Some firms had difficulties in integrating their legacy systems into the new, largely web-based systems that had been developed to capture APIS data:

> The system that we use, when it hits a weekend, the amount of new bookings that get made is usually higher than in the week. When there are so many new bookings made, it doesn't filter in the system properly. So if someone has made their booking on Friday, it doesn't go in ... so they're trying to put their passport information in the next day and they can't do it ...
>
> (T12, Call Centre Team Leader, Tour Operator)

Travel firms thus became concerned that eBorders would sour the customer relationship, first because of the delays described in the previous extract above, but also because of the mandatory nature of data collection. Interviewees were worried that customers may have been prevented from travelling if they had not provided the necessary APIS data. As this interviewee commented, it is hard to judge how to manage that situation:

> It's trying to get the balance between getting your customers to provide the data and the responsibility that you have for collecting it. What happens if they don't give you the data? Do you come down on the customer quite heavy-handed? Do you withhold tickets? That's a really difficult thing ...
>
> (T4, Business Development Executive, Airline)

By contrast, financial services interviewees reported that cost and contextual tensions arose between the firm and its customers. AML/CTF caused interviewees to identify that disproportionate costs were associated with regulatory compliance. As one interviewee explained, this reduced expenditure on customer-facing activities:

> There is more time and money being spent on overheads, such as regulation and AML, than on sales.
>
> (F3, Business Delivery Manager, Retail Bank)

Financial services interviewees also felt that delaying or denying a financial transaction, as a result of reporting suspicious activity, was in tension with their commercial pressure to provide prompt, reliable service to customers. It is illegal for the financial services provider to inform the customer that they have been reported or explain why the transaction will not be completed. This accentuates the apparent lack of customer service and has the potential to damage customer relationships:

> In the corporate world, if a client wants to transfer £50 million to another organisation, their bank has to apply to NCA ... There is roughly an 87 % approval rating ... Banks must tell their clients that they are not in a position to perform the transaction, but not why. As some organisations have found, if they start legal action against the bank they will then realise that they have been reported for suspicious activity.
>
> (F5, MLRO, Investment Bank)

Tensions between supply chain partners

While financial service sector interviewees did not report any tensions with supply chain partners arising from AML/CTF, the opposite was true for the travel sector. Many airlines and their supply chains, i.e. tour operators and travel agents, collected APIS data by developing customer 'self-service' web sites so that customers could input their passport information themselves. These websites were not only used by customers, they were also used by travel agents who inputted data on the customer's behalf. As a result, agents who used a range of airlines were faced with a range of different interfaces, time limits and instructions. Whilst many of these systems worked well in their own right, developing interfaces that balanced customers' wishes to input their data at different times, technical limitations as to when downloads and updates can be carried out and the rigid time window placed on the transfer of data by UKBA was a challenge. The myriad of interfaces available also placed commercial pressure on travel agents and tour operators. They felt that they had to input APIS data for the customer to prevent them going to the websites of the airline or tour operator with which they were travelling and finding better deals there. This tour operator explains:

> The initial mind set of an online travel agency was we've made the booking, our job is done. Then you realise that you need to retain customers ... are we prepared to take the risk of pushing the customer directly to the airlines ... we need to keep the relationship in-house, get them to submit the information to us through some sort of portal interface.
>
> (T28, Customer Operations Director, Travel Agent)

For a short time, the Association of British Travel Agents considered developing their own APIS data capture site, to prevent this loss of customers:

> Our problem with someone like [tour operator name], was, we don't really want our customers going onto their main website, putting all of their data into the website, just to meet the [eBorders] obligations; because ... they are getting one step closer to becoming direct customers as opposed to our customer. It's the retail travel agent's paranoia but it's a valid paranoia.
>
> (T21, Commercial Director, Travel Agent)

Smaller agents and tour operators felt particularly disadvantaged because they did not have the resources to develop their own information systems which stored customer data for future use. As one tour operator explained:

> This has implications on the customer interface side: because customer information is destroyed when the flight leaves, customers who repeat purchase have to re-enter their information. The difficulty with investing in a full CRM or frequent flyer system is that travel agents margins are so small they do not have the money to invest in these systems.
>
> (T8, Call Centre Manager, Tour Operator)

Tensions between competitors

Tensions between competitors in the travel sector arose largely due to the multiple systems and the choice of the data transfer protocol, discussed earlier. Interviewees described how the existing systems favoured some carriers, particularly the larger, legacy carriers who were part of global alliances. While the larger legacy airlines used Global Distribution Systems to collect and transfer APIS data, as this interviewee explained, smaller retail carriers had to find novel ways to address the new requirements:

> For small carriers, e.g. small airlines on Greek Islands – they could buy a scanner on the Internet that could read the data on passports and transmit this over the Internet to the UK.
>
> (T3, Assistant Director, Government Agency)

Interviewees in the travel sector described how the eBorders regulations were starting to influence the service level agreements between tour operators and airlines and hence were becoming a point of differentiation between competing airlines, as described by an airline employee:

> Particularly in the US where it's even more stringent, we have to provide the data 72 hours in advance, we are having to write that into contracts [with tour operators] ... But we don't want to do that because if another airline doesn't write that into their contract, [an operator] could find that they've got better terms going somewhere else ...
>
> (T4, Business Development Executive, Airline)

Interviewees from the travel sector also highlighted that competitors were likely to adopt different approaches to the care they took in collecting data. This would be likely to reduce the costs they incur whilst reflecting badly on the whole sector.

> I know that certain carriers, from certain airports, will choose to do a shoddy job because it's easier to do a shoddy job ... eBorders don't know that whether the 150 names collected online was the total on board the aeroplane or whether there was another ten.
>
> (T4, Business Development Executive, Airline)

This lack of care represents an unintended consequence in the form of a type of resistance. It is in contrast to the sentiments expressed in the financial services sector where individuals and organisations were keen to highlight the care and attention that was placed on AML/CTF matters. This difference appears to arise, in part, due to the responsibility and the liability that individuals face if they do not comply with AML/CTF, which will be discussed further in the next section.

The greater maturity of AML/CTF and the lack of need to adopt imposed technical standards meant that few technical issues occurred. However, interviewees in the financial service sector noted that the significant costs involved in meeting the regulations resulted in smaller organisations finding it more difficult to support the necessary investments in systems, staff and training:

> Basically, the bigger the organisation the more time is spent on AML. For example, [name of large bank] probably have a team of 2,000 working on it. Smaller organisations probably are busy with the everyday running of things and may not have a staff member who concentrates solely on AML.
>
> (F5, MLRO, Investment Bank)

Tensions between employees and their organisation

The final tension resulting from these regimes arose between the employees and their employers. In both sectors, employees, particularly those on the front-line, had to accommodate regulatory responsibilities into their everyday jobs and balance the competing demands of their employers, customers and the government. Tensions, indeed anxieties, arose in relation to technical, cost related and contextual matters.

eBorders compliance has not been met by significant increases in staff. Hence its impact on existing roles is significant for the travel firms which tend to be smaller and have lower operating margins. Interviewees in one tour operator, which had developed its own APIS data capture website, had 30 staff in their call centre working full time on supporting the queries from customers who were trying to provide their data via the site. Interviewees in the travel sector, particularly those in customer-facing roles also described how compliance with eBorders had changed their roles in ways that were in tension with their customer service role. For some employees, eBorders data collection had become a significant, but unwelcome, part of their job. Staff interviewed in the call centre of a tour operator described how they found the calls repetitive and boring and how they reduced their opportunities to meet their sales targets. Similarly, the importance of compliance with the eBorders scheme resulted in the unintended consequence of employees in the travel sector feeling anxious:

> It's nearly doubling up your workload. Maybe we're more paranoid than most but I would have a look at on-going stuff on our computer system maybe for the next month and say, there's five people going here and going there can we just double check that all of that information is in.
>
> (T16, General Manager, Travel Agent)

Whilst financial service firms make significant use of IS to identify unusual patterns in transactions, it was observed that monitoring relied on a combination of IS and vigilance by staff:

> Anti-money laundering, partly because of our size and partly because of the nature of our business, it's not an easy thing to fully automate ... and so we have a mixture of system support, which probably only forms 30 % or 40 % of our controls, and the rest of it is due diligence by the staff.
>
> (F19, MLRO, Investment Bank)

The regulations effectively make all employees responsible for identifying and reporting suspicious financial transactions. Indeed larger banks had significant numbers of specialised staff working on AML/CTF issues as well as training front-line staff to spot suspicious activity when dealing with customers on the phone or face to face. This did not always sit well with staff. The following interviewee, from an investment bank, notes how difficult it was for traders to spot suspicious activities:

> They [staff] are focussed on doing deals and have very short-term memories – they can't even remember in the afternoon, the details of the deals they did in the morning.
>
> (F6, MLRO, Investment Bank)

If staff failed to spot something they were threatened with fines and imprisonment. Unsurprisingly, high levels of anxiety around the regulations resulted. Concerns about how such failure might jeopardise the future of organisations are reflected in the comments of one member of staff in the financial services sector who referred to the fact that:

> A bank in the US was brought down because of their lax in AML controls – it literally brought them down.
>
> (F18, MLRO, Retail Bank)

Table 4.1 summarises the tensions which arose in both the financial services and travel sectors as a result of their mandatory participation in AML/CTF and eBorders respectively. This summary highlights the main difference between financial services and travel sector organisations. The more established nature of AML/CTF, as well as its requirements for financial service organisations to analyse and to submit summary reports, rather than to simply collect data from extended supply chains and then transmit to the regulator can begin to explain the tension patterns.

Conclusions: Tensions, disruption and realignment

In this chapter we have drawn on the theoretical lens of information infrastructures (IIs) to begin to break open how two governmental surveillance regimes which collect consumer data for national security purposes enmesh with the activities of private sector organisations. IIs have allowed us to explore how the different types of tension which arose between the stakeholders involved due to technical arrangements (Henfridsson and Bygstad, 2013), cost (Edwards et al., 2009) and context (Ure et al., 2009).

The central premise of both of these schemes is that they transform private sector infrastructures into information conduits for national security. Customers, employees, supply chain partners and competitors of focal firms were each affected by the way in which the regime disrupted their existing stakeholder interests via infrastructure. A range of physical and social infrastructural objects whose primary aim is to serve the interests of the private firm were then implicated.

In the travel sector, the UKBA saw a single inflow of passenger data from airlines making journeys into and out of the UK as the most efficient way to make the same translation. Internal information systems of tour

		Stakeholders				
Nature of Tensions	**Sector**	**With Customers**	**With Government /Regulator**	**With Supply Chain Partners**	**With Competitors**	**With Employees**
Technical	**FS**	*Not mentioned because customers are not aware of AML/CTF investigations during the service encounter*	*Not mentioned; SARs are transmitted to NCA through a well functioning portal*	*Not relevant because AML/CTF does not call supply chain relationships into question*	*Not mentioned, because AML/CTF concerns internal technical adaptations*	Very difficult to fully automate – requires systems and human vigilance
	Travel	Integration with legacy systems Managing uneven demand (weekend peaks)	Influence of system suppliers has led to adoption of less used standards	Inconsistent approach especially to system interfaces – causes additional work for partners	Multiple systems used across travel sector and perception of advantage of bigger players	Some cases staff willing to 'stand in' for customers as work around for system limitations
Cost	**FS**	Costs of compliance increase overheads and reduce amounts spent on customer-facing activities	Cost of on-going regulatory compliance – systems operation and staff training	*Not relevant because AML/CTF does not call supply chain relationships into question*	Smaller firms cannot afford dedicated staff or units	Has led to new roles and, in larger firms, new units. All staff require on-going training and monitoring of their AML knowledge and performance

	Travel	Costs of initial development of systems and on-going systems operation and staff training	Centralised system imposed on travel sector viewed as high cost	Concern about margins in travel sector preventing smaller partners developing solutions	Disproportionate extra cost on smaller players as need to collect new information	Have had to provide staff time to collect APIS data – sometimes dedicated roles and sometime as part of existing roles
Contextual	**FS**	Balance customer service and AML obligations – cannot discuss with customer as 'tipping off' is an offence. Carry out repetitive screening	Government has made training mandatory – leads to high awareness in focal firms	*Not relevant*	Severe sanctions lead to consistent approach across FS firms	AML role clashes with selling, service or deal-making role. Led to an atmosphere of anxiety and fear
	Travel	Additional time taken in service encounter. Difficult to determine how 'heavy handed' to be with customers. Collect extra data 'just in case'. Little public awareness	Government has not made training mandatory – leads to piecemeal and varied levels of training and awareness in focal firms. Firms in sector feel Government has not listened to them – rather has been driven by system suppliers	Small tour operators and travel agents very concerned about damage to customer relationship	Seriousness with which firms implement eBorders can act as a differentiator between competitors	Has made some roles repetitive and boring and reduces opportunities to meet sales targets

Table 4.1. Tensions associated with the II schemes.

operators, travel agents and ground handling agents in airports at home and overseas became extensions to the security infrastructure. Organisations in the travel sector were required to share passport information across their supply chains and with the UKBA using a non-standard protocol. Information was transferred to the UKBA en masse and no analysis of the data was done locally by firms. Aligning different legacy booking systems within and between organisations and with airport and ground handling systems became the main infrastructural development activity. Fines for non-compliance and a decentralised approach to training and data collection practices were the ways in which this infrastructure was governed. Significant tensions arose between all stakeholders concerning technical and cost issues, while contextual questions over the legality of the eBorders regime and unclear training requirements arose. Rather than practising vigilance, travel sector firms spent their time devising work-arounds, forging connections and adapting their practices as the whole II evolved. The varying degrees of centralisation within the regimes afforded participating firms differing degrees of autonomy in how they responded, demonstrating the paradox of control that has been associated with large IIs (Hepsø et al., 2009). The centralised mass-data collection pursued in the travel sector meant significantly more adaptation had to be made within firms and their supply chains. A range of interfaces in numerous corporate locations ensured that the information flowed from customer to government.

For the financial services sector, AML/CTF prompted the continued investment of financial resources to support training infrastructures so that staff would be able to identify suspicious activity to create the required information flows for the scheme. Service and sales practices were challenged by an infrastructure which attempted to translate the customer into a threat to the nation rather than an opportunity for the banks. Due to the more established nature of AML/CTF and the prevalence of local data analysis evidence of a more stable infrastructure emerged in financial services firms. Fewer technical tensions were reported by the financial services interviewees. Greater choice was given to firms as to the identification of risk and the manner of their response, but this resulted in reports of varying quality (Ribes et al., 2013).

As well as these obvious differences there were some similarities between the sectors. A pattern emerged in relation to industry structure, particularly firm size, of the ability to comply with the regulations. Smaller firms found it difficult to accommodate the regulatory demands and did not have the funds or capacity to develop information systems to expedite matters. It also became clear that firms in both sectors experienced similar contextual tensions as the competing interests of national security with internal aspects of the firms operations were reconciled. In both sectors this related first of all to the management of the customer relationship and to how staff incorporated regulatory requirements into their everyday jobs.

Our exploration of tensions between stakeholders shows that the number, range of types and widespread distribution of tensions across stakeholders indicates the depth of the regimes' commercial impact. The tensions which arose primarily concerned the perceived disruptions caused by the regulation. A number of themes arose in the data which indicated the disrupting effect of the regulations for firms and their employees. This can be illustrated by expressing these themes as a set of competing pressures in which issues of compliance and issues of competition can be seen to be pulling organisational processes in different directions. These competing pressures indicate the qualitative nature of the tensions experienced, for example:

- Financial pressures which demand money be spend on compliance rather than on sales, commercial and customer service priorities.
- Pressure arising from organisational size and business relationships, concerning the proportionate impact of the regulations on firms of different sizes and differing positions in the supply chain.
- Pressure arising from compromise, where firms perceive a decline in their autonomy in defining commercial strategies as well as a threat of impact on corporate reputation.
- Pressure arising from the customer relationship where firms have to balance the necessity of gathering security information from customers, which constructs them as a threat, as well as maintaining a good customer relationship and managing the workload of customer-facing staff.
- Pressure arising from influence at governmental level, where the content of government consultation with firms over eBorders was not carried through into a set of outcomes relating to system design and standards.

The nature of these pressures gets to the heart of the notion that converting private sector firms into national security data conduits is fraught with problems of realignment between resources in the firm. Capital, labour, systems, intangible assets such as reputation and brand are all implicated. These are resources which already constitute existing organisational processes that pursue particular institutional ends, but are then mandatorily subsumed by a higher governmental priority on pain of criminalisation. The language used by interviewees to describe its effect is one of inconvenience and compromise – to use the language of stakeholder theory – to their own mission, legitimacy and power. It pitches the perceived autonomy of the firm and its ability to use its resources as it wishes against a governmental claim to use those resources for its own ends. Issues of rising cost, changes to the terms of engagement with the market and competitors, threats to service and reputation were expressed by the range of informants we interviewed. Destabilisation in these core components of the 'business of business' encapsulate how dealing with mandatory security regulations as a feature of regulatory responsibility is potentially disempowering for the firm. In the next chapter we examine how firms renegotiated their involvement in the schemes by realigning internal resources in such a way so that commercial priorities were re-established. The public interest of security slowly became subject to the market logics of the private sector firm. This time, we consider eBorders and AML/CTF using theoretical tools from the business disciplines of strategy and marketing.

CHAPTER FIVE

The strategic response

Recognising, rationalising and refashioning in the retail travel customer relationship

Introduction

In the next two chapters we consider the response of the private sector organisations involved in the eBorders and AML/CTF schemes. At a strategic level of analysis, we focus on how these firms absorb the requirements into their day to day activities, the actions they take to maintain their commercial interests, and the implications for their marketing and customer management activities. We offer further insight into the internal resources which are mobilised around new securitised information flows as well as the rationales which pursue the marketisation of security activities. As the implementation of eBorders is at a much earlier stage than the well-established arrangements for AML/CTF, there are implications for the extent to which the processes required by the schemes have been embedded in the organising processes of those firms who have to comply with the schemes. Marketing and customer management are important sites to examine within the organisation, because the point of customer contact for commercial purposes becomes the point of customer contact for security purposes under the regimes. In the case of AML/CTF the processes and systems for handling the required data monitoring tend to be highly integrated with regular customer relationship management approaches used by banks and other financial services providers. They have become absorbed into day to day working practices surrounding customer service. The situation for travel firms implicated by the eBorders' requirements is more emergent, as firms build their understanding of compliance and what it means for their commercial interests.

Chapters 5 and 6 focus on two main questions which relate to how the private organisations involved in these partnerships have changed their practices to fulfil the requirements and the ways in which they have maintained their commercial interests:

- What actions are taken to protect the commercial interests of firms within these public-private security arrangements?
- How are the requirements placed upon these firms absorbed into their day to day customer management practices?

The analysis of our data focuses on the marketing responses of firms and on how practices change to reflect the need for compliance. We show how the regulations become subject to the internal market logics of firms. In the retail travel case, this relates to how the firms respond to the changes eBorders brings to the macro setting in which they operate and how those are incorporated into existing practices. Market logics emerge which relate to firms taking ownership of the customer relationship and adding value to it by cross-selling. In the retail financial services case, the focus is on how AML/CTF requirements have become integrated into customer management practices. Market logics emerge which relate to the value of the 'compliant sale' where both regulatory and commercial objectives are met. In each case these logics represent the re-combinations and re-orderings which take place as a result of the operational and strategic disruptions mounted by the regulations. In this chapter, we begin by considering the emerging response of the travel sector to eBorders, before moving on to examine how AML/CTF has become established in financial services firms in chapter 6.

Introduction: Mobilising the customer relationship to enact security measures

As explained in Chapter 2, the eBorders programme involves all air carriers collecting and electronically transmitting travel document information and service information for all passengers travelling out of the UK. An additional layer of complexity for carriers is that they are also required to collect data from firms that sell aircraft seats on their behalf. Consequently, tour operators, travel agents and seat brokers are all affected by eBorders and must develop systems and processes for managing the data gathering and transfer. In Chapter 3 we outlined how each group involved in surveillance-based national security regulation has particular interests which influence their response to compliance requirements. We outlined how these interests sometimes conflict and are experienced as tensions in the sectors. Several key tensions arose in eBorders between governmental and commercial interests. The UKBA's remit is to manage immigration and border security under the eBorders regime using data gathered by companies, but it is not concerned with protecting commercial interests in the travel sector. Customers, on the other hand, are seeking easy and efficient travel from location to location, but in order to do so they must provide their passport data prior to travel. Airlines and travel agents meanwhile must maintain good customer service so that customers can achieve their desired travel experience and they can protect their commercial position, but must do so while complying with eBorders' requirements.

In Chapter 3 we also highlighted how the stakeholder relationships, which emerge as a result of the regulation, are interconnected across different stakeholder categories as well as with the focal firm. A feature of eBorders is that the actions of each stakeholder have implications for the others, but that the combined influence of the stakeholder group on the carriers is greater than each acting separately. Within retail travel new forms of business relationship needed to emerge in the air carrier's supply chain so that data could be transferred. While the air carrier is ultimately responsible for the gathering and transfer of passport data, organisations who are selling air seats in its downstream supply chain – such as the travel agents and seat brokers – are the ones who have the direct contact with customers and so they need to gather the information so that it can be sent to the airlines. Air carriers are therefore also responsible for ensuring that other travel firms pass on the information that they need. These new forms of connectivity between supply chain partners have commercial implications to which all firms need to adapt.

Collaborative processes and long term outcomes

When stakeholder interconnectedness is viewed through the lens of stakeholder collaboration, a number of insights emerge. Evidence from tourism research suggests that collaborative projects are more likely to achieve their objectives when decision making is based on consensus, information is widely shared and partners' involvement evolves over time to reflect the collaboration (Arnaboldi and Spiller, 2011; Jamal and Getz, 1995, Vernon et al., 2005). Relatively little consideration has been given in the literature to the long-term outcomes of stakeholder initiatives. Understanding how stakeholder relationships change over time, including how the commercial performance and customer relationships of firms are affected, is crucial to understanding the success or failure of collaboration (Friedman and Miles, 2002). Evidence suggests that the significance of particular stakeholders in a partnership changes over time as new alliances are formed (Mitchell et al., 1997) and the environment changes (Phillips, 2003). Examples of such changes in relation to eBorders include delay of the launch, the dismissing of the main IT contractor (Kollewe, 2010) and the EU legal challenges, all of which have created uncertainty in retail travel. For example, the long-term outcomes of a particular collaboration are heavily influenced by how the partnership is structured and the extent to which local actors are involved (Arnaboldi and Spiller, 2011).

Travel firms found that eBorders compliance conflicted with their commercial interests. It was costly and operationally disruptive, primarily affected employee work patterns and morale and negatively impacted the interactions with customers and the service provided. For example, the collection of required data alone is time-consuming and adds 'at least 15 minutes to the customer service encounter, no matter what medium' (T12, Airline). Staff often found themselves helping to in-put customers' API data, even though their organisation had created websites which should have allowed customers to do this for themselves. In the words of one interviewee:

> I have had some customers that refuse: they don't want to bring us their passports at all, in which case I then put the onus back onto them, and it is their responsibility. And if they haven't done it, then they can't come back to us.
>
> (T22, Tour Operator)

Others were concerned about protecting their firm's customer data from competitors and felt that their ownership of the customer relationship might be threatened. For this reason, some interviewees were reluctant for their customers to provide their data via the carrier's website:

> If you're a cynical travel agent like we are, you might say, well, hang on a moment; if they fill in all that information … what's to stop you using that data for other marketing activities.
>
> (T21, Tour Operator)

'To compete or to collaborate?' was the question the retail travel sector faced when considering their initial operational response to eBorders.

Creating the information flow: The impulse to collaborate

Non-competitive collaboration between airlines and their supply chains emerged as the initial strategic response to the eBorders regulations. In Chapter 2 we described how, as a precursor to eBorders and before it appointed the Trusted Borders consortium, the UKBA set up a Regulatory Impact Assessment consultation with the travel industry and the initial collaborative stance emerged from that process. Considerable discussions between stakeholders were required as the UKBA consulted airlines, tour operators, travel agents and industry bodies such as the Association of

British Travel Agents and the Airlines Industry Group. The standard for the gathering of passenger data was set during these discussions and the consortium was appointed. In order to keep data collection away from 'the travel experience' and to meet the data standards of the UKBA, airlines attempted to capture data 'upstream' i.e. before the customer arrived at the airport to travel. The following extract from an interviewee in a travel industry association describes the initial difficulties posed to the sector by eBorders:

> They capture data at the airport at the moment, but eBorders means that it has to be captured upstream. Also they don't have automated check-ins in many countries and so the onus falls on the airlines to do it for them. They can't rely on every airport to have the ability to scan passports and send the data on. All of the UK airlines need to have the ability to do this by the end of this year. The tour operators and their carriers are having to alter the terms and conditions of carriage in their brochures to allow for the fact that some people might not have provided their data and without it they can't take them!
>
> (T12, Trade Body)

This interviewee describes how the tight time frame for transmission of passport data to the UKBA initially encouraged all air carriers to try to capture passport data well in advance of travel. Charter carriers were particularly affected because they sub-contracted their check-in procedures to ground handling agents and were subject to extra charges per passenger if their passport needed to be swiped at check-in. The focus on a standard format and timing of data capture prompted many of our interviewees to point out how the design of the scheme did not take account of the diverse modes of business in the sector, particularly in terms of how easy it was to contact customers for their data. They highlighted conflicts which emerged at a strategic and commercial level. They first took issue with how pre-emptive data capture challenged established business models and information transfer protocols dictated by the scheme to be at odds with, more expensive than, well established industry standards. One compliance officer refers to the almost overt conflict with UKBA which emerged during the eBorders negotiations, relying on the notion of the 'business model' to resist the changes eBorders were demanding to interfaces and transmission protocols. The following quote highlights the rigidity of eBorders and the industry stand-off as they tried to hold their commercial ground:

> Yes, so there's a rigidity; there's no flexibility there ... Of course, when we come to start talking about implementation all of these worms crawl out of the woodwork. And they're not worms; they're massive anacondas ... the interfaces and processes just didn't support our operations. And we were pressurised into moving in a direction that didn't support our business and we rebelled ... We had very heated discussions on this to a point where the government suggested that we might have to change our business model because of it. That's completely unacceptable.
>
> (T6, Airline)

Inflexibility was further highlighted as organisational differences between national legacy carriers, such as BA or KLM, budget airlines such as Ryanair and leisure or charter carriers, such as Monarch, emerged. Each type of carrier developed a response which suited their own style of business. KLM and BA, with huge frequent flyer programmes and proprietary airport infrastructures found it relatively easy to capture data upstream to aid UKBAs pre-emptive data collection. Many budget airlines, however, sought to keep costs down and moved to 100 % online check-in to avoid additional airport costs. They refused travel to passengers who had not checked in online. Charter carriers tried to maintain genial customer and commercial relationships, for fear of losing customers to competitors. They gently tried to persuade customers and the travel agents and tour operators who sold seats on their aircraft, to capture data upstream. As one interviewee explained:

> In terms of our agents, we talk to them about the need for them to help us with the collection of data... if you force them against their will they'll just put something there that's meaningless and that means that we've got to do it again at the airport.
>
> (T6, Airline)

Amongst the airlines there was a sense that they alone were making eBorders work and they were critical of the role of other carriers as expressed by these two interviewees, the first complaining about the haphazard approach taken by low cost airlines and the second about the response of non-air carriers:

> I know that certain carriers ... will choose to do a shoddy job ... The [Airline Name] ... for example, all the stuff they collect online, they can send fairly easily. But anything where the passengers don't complete the web check-in process ... they have to collect manually at the airport. So they might go, this manual bit's a bit difficult; I won't bother.
>
> (T4, Airline)

> I think you'll find that the ferries certainly, and possibly Eurostar, are doing very little, although they represent a big chunk of what comes through the border. So, there's an impact and competitive situation there that annoys us a bit.
>
> (T15, Trade Body)

This is significant because it was widely perceived that UKBA assumed that because national legacy carriers would find it easy to comply, so would everyone else. In reality, compliance required significant investment on the part of the diverse air carrier sector not only in systems and practices but also in the maintenance of their business relationships to ensure that their business models were not compromised. One IT manager in retail travel commented 'they just don't understand the charter airline process'.

The practicalities of the information flow: The long term view

eBorders presented a change in the macro-level operational context for carriers and their supply chains to which they then had to respond at a strategic level. When eBorders was initially implemented, carriers, tour operators and travel agents focused on complying with the requirements but in minimising its impact on their commercial priorities. Good working relationships formed the basis of a strong response from key players in the charter sector. In the face of the eBorders programme, which was perceived by the sector to be an unfair regulatory burden and to contain conflicting messages, certain key players in the charter sector presented a united front to regulators, travel agents and tour operators. As this airline IT executive explains:

> What we then decided to do was work as part of the charter airline group with Airline 1, Airline 2, Airline 3 and ourselves and we tried to keep it as a group, so there wouldn't be competitive advantage or disadvantage ... so we each have our area of responsibility which helps cover the cost better and from an industry point of view because we're airlines we have a number of tour operators out there ... what we wanted to do is give a message to them that each of the airlines are asking for this in this format which is standard because what we didn't want was 'Airline 1 said just send us this information' and then they'd have a competitive advantage over us because they would use them as an airline rather than us. So if we have a consistent message and we all have to do it for regulatory purposes [it] works really well.
>
> (T7, Airline)

The principle of this strategy was that key industry players agreed between them that they would not want to give away competitive advantage on a regulatory matter. Each player agreed that they would build their own upstream data capture micro-site, to which customers would be directed after they booked. eBorders compliance would be the responsibility of the customer, companies would not get involved and the whole thing would have a minimal impact on their operations. They also agreed that no advertising would be placed on the micro sites, nor would they sell on the data collected from customers if they called the carrier, operator or agent with an eBorders related query. This did not stop internal marketing departments being tempted by the opportunities which arose from having another point of customer contact, but as the micro-sites did not capture live e-mail addresses of customers they quickly lost interest.

In spite of their efforts, gaps still emerged in the customer data flow. For example, the charter sector only managed to capture 35–40 % of customer data upstream as opposed to 98 % reported by some legacy airlines who could use the Global Distribution Systems to transfer data. However, more deep rooted issues connected with the customer relationship management in retail travel began to render the non-competitive 'pact' precarious, as the carriers worked to make this a reality. Issues arose with customer booking practices, customer familiarity with the eBorders processes and internal system lags.

The first problem concerned the way in which eBorders disrupted everyday practices of booking retail travel. Inserting a requirement for all customers to provide passport information at the point of booking may have been simple for frequent flyers with accounts at BA or KLM, but it is a challenge for the high street travel agent or tour operator selling package holidays and flights. Numerous issues arise. The first is that many holidaymakers do not go to the travel agent with their passport, nor do they expect to give their passport data for regulatory purposes:

> I don't want to ... confuse the customer, thinking oh God, I've got to give you this information ... we specialize quite a lot in golf groups, so a group of 12 lads going down to Portugal to play golf. They don't know the passport details for John that they've met once, who's a friend of a friend of a friend who just happens to be coming ...
>
> (T4, Airline)

More experienced travellers, who may be frequent flyers with a legacy carrier, object to providing data repeatedly for carriers with whom they are not frequent flyers. An IT executive at a travel agent noted the sensitive way in which this had to be handled, with the different points of sale integrated into the communication strategy around eBorders. Whilst lamenting the lack of media and government publicity surrounding the need to provide passport information, they are careful to avoid threatening the customer with regulatory requirements such as the denial of travel or charges if they do not provide data:

> We have web, travel shops, call centres so with each of those how do we inform the customer that they have booked obligation and I think one of the disadvantages we've had is that there has been no real press coverage of all this to the normal customer.
>
> (T23, Travel Agent)

A further problem emanates from time lags within tour operators' internal systems and their transfer of data to the carrier. Many people book package holidays 18 months in advance, yet the booking data only becomes live in the carrier's system 6 months before departure. Hence the carrier is then tasked with re-engaging the customer at that point and asking them to provide their information, long after the sales encounter. One carrier also reported considerable investment in a user interface for self check-in kiosks to avoid problems at check-in.

Once the operational and infrastructural challenges had been addressed, the need to protect customer relationships and to minimise the impact on service came to the fore because of the commercial threats posed to that relationship by the regulation. These concerns were initially manifested in the actions of firms to protect their customer relationships and subsequently in their efforts to explore commercial opportunities that arose out of the initiative. As the following quote reveals, achieving a balance between the need for regulatory compliance and the protection of commercial interests, was not easy:

> ... it's trying to get the balance between getting your customers and bringing the booking in and then the responsibility that you have there, and then collecting this data. What happens if they don't give you the data? Do you come down on the customer quite heavy-handed? Do you withhold tickets?
>
> (T4, Airline)

Firms were concerned that information gathering did not conflict with the need to provide good quality customer service. Frontline staff were also keen to do what was needed to ensure that customers were able to travel. Rather than using the travellers' obligations to provide data as a 'big stick', they reported doing what was necessary to support customers as they provided their data. One interviewee spoke of the perils:

> if you don't get the right details in and you've got a digit wrong, what the effect could be on your customer who's at the airport and it's your fault.
>
> (T19, Travel Agent)

As such there was a clear shift in the way in which the retail travel industry addressed eBorders as its full implications for the customer relationship became clear. At the start of the roll out, travel firms viewed the inputting of data as the customer's problem, but more recently, firms have been keener to take ownership of the activity. Not all customers were able to enter their data properly and there were concerns that customers might decide in future to make all of their own travel arrangements. One customer manager explains the shift in thinking:

> It was, push it on the customer and we're not worried. And it was like we've made the booking, our job is done. We need it to be low cost; we don't want to speak to that customer again if we don't have to ... Then you grow and you realise that you need to retain customers, we need to nurture relationships ...
>
> (T28, Operations Director)

Under eBorders' rules, those we interviewed knew they were not allowed to exploit passport data for marketing purposes. Some were doubtful as to its commercial value while others suggested that their sector anyway less used to using aggressive marketing methods than some others. Those interviewed did, however, acknowledge that the level of customer interaction has increased and that eBorders offers '... a good excuse to get in touch with your customers again' (Airline T4). But this benefit was tempered with concerns about what eBorders meant for the ownership of customer relationships. There was evidence of turf wars between rival travel businesses, as one travel agent explained:

> so our problem, immediately, with someone like Thomas Cook, was, we don't really want our customers going onto your main website, putting all of their data into you website, just to meet the obligations; because in filling in that e-form on your website, actually they are getting one step closer to becoming direct customers of you, as opposed to our customer ... It's the retail travel agent's paranoia but it's a valid paranoia.
>
> (T21, Travel Agent)

Such concerns about ownership of the customer relationship are now shaping the industry's response to eBorders, as is the desire to explore commercial opportunities arising from the extra customer contact which it brings. In particular, even though firms cannot directly use passport data for marketing purposes, they increasingly see eBorders calls as an opportunity to improve existing bookings and to sell 'extras' such as the choice of seat or meal on flights. As one frontline team leader put it, 'A call is a call at the end of the day' and 'I try and tell them in their call coaching sessions to sell ... try and sell them something' (T13, Tour Operator). Such cross-selling is already common practice in the industry, so these extra selling opportunities are generally welcomed:

> Historically, you go into travel agents and you would buy a ticket, the tickets would turn up in the post two weeks before you travel. You would never speak to a travel agent, never speak to your tour operator and never speak to your airline until you get to the check-in desk. And that was it. But now passengers who book direct, can you give us your APIS (passport) details, and while you're doing it, do you fancy buying a leg-room seat? Do you fancy buying car hire? Do you want to buy some insurance? Great cross sale opportunity, absolutely fantastic cross sale opportunity.
>
> (T4, Airline)

While the full marketing potential of eBorders may yet to be realised, the opportunities to build customer relationships and generate extra sales income are recognised and increasingly being explored. The development of 'known traveller' approaches, whereby the customer shares certain personal data with the travel firm, in exchange for a smoother process when they travel, is one example.

eBorders and the customer relationship in retail travel: The emerging market logics

The story of the embedding of eBorders' requirements within the retail travel sector is a story about organisations attempting to gather data from customers with minimal impact on the customer relationship. It was revealed that in taking such a stance, there were commercial implications for the firms concerned and they responded accordingly. The initial response was to keep eBorders away from the competitive process of selling travel products. But the amount of customer contact it precipitated then prompted companies to think of ways in which to capitalise on these additional contact points. The point of sale became the point of security data capture, but that moment of contact around security data capture then became re-subsumed in the commercial processes of selling and service delivery. The retail travel organisations we investigated formally refashioned their response to the regulation having rationalised its impact on their businesses (Dibb et al., 2014). We discuss each of these processes in turn.

Recognising

In the early stages of eBorders, travel industry stakeholders learned about the rules of engagement and the decision making that would take place. This then established the foundations for the distribution of power between the different parties. At that point, the stakeholders were engaged in a process of *recognising* the requirements and likely implications of compliance. There were discussions between stakeholders about how the requirements could be met and what the impact would be for operational processes and ways of working. A range of intangible costs, including transaction time, complexity of the process, customer confusion and the possibility of mistakes, were seen to impact upon service delivery. The fact that the customer interaction was negatively affected was a major concern in a sector that emphasises service quality and the customer experience (Shaw, Bailey and Williams, 2011).

Rationalising

During the *rationalising* stage, eBorders compliance requirements became aligned with operational factors, such as the modification of technology, processes and ways of working. Rationalisation either followed or took place alongside the recognising stage. The evidence suggests that travel firms found this process to be costly and difficult and that the relationships between stakeholders can become destabilised, as different players moved to protect their commercial interests. Not surprisingly, the necessary collaborations between firms were affected by a number of tensions and uncertainties, which negatively impacted upon customer relationships. We found the relationships between the carriers and the government and IT supplier consortium to be particularly problematic, as is often the case when power brokers disproportionately influence the process (Jamal and Getz, 1995).

The relationship between the carrier, as the focal point in the network and their agents and partners was crucial. It was these firms which were directly in contact with customers and which needed to implement the gathering of travel data. These firms also, bore the costs of implementing new systems, were disrupted by putting new processes in place and had to train and recruit front-line workers. Developing strong collaborative networks between these organisations, even though some were direct competitors and smaller operators may have little power in the process, was likely to benefit all parties. Together, these stakeholders were more powerful and had greater influence over the government and technology suppliers and also, therefore, greater control over the technical solution which was implemented.

Refashioning

Carriers and other travel organisations were required to modify their working practices in response to the demands of eBorders. We use the term *refashioning* to describe these changes to their ways of working. During the refashioning phase organisations explored a variety of ways to limit the costly, and at times, destabilising aspects of compliance. Ways of delivering high quality customer services within the regulatory framework and new approaches to develop commercial advantage were considered. While many of these refashioning activities accompany the rationalising phase, some began earlier. There were two main drivers for this refashioning phase, which is where market logics emerged. Firstly, customer relationships needed to be nurtured carefully and managed, to ensure they were not threatened by compliance. Although carriers initially tried to shun eBorders costs by passing them on to travellers, the mood quickly changed amidst angst about customer retention, to easing the burden of compliance for customers. Secondly, despite an initial suggestion that UKBA rulings must not be a source of commercial advantage for firms, as time passed more such opportunities were sought. While in the early stages of implementation we found little evidence eBorders being used to generate commercial opportunity, in the later data gathering organisations were increasingly using the customer contact required by eBorders to cross sell other products and services. We anticipate that such actions are likely to become increasingly routinised over time.

Conclusion: Market logics, commercial priorities and the customer relationship

Like AML/CTF, eBorders is a regulatory framework with which commercial firms must comply and which involves partnering between public and private organisations to ensure the flow of security information back to government. In this chapter we have considered two questions relating to the actions that travel firms take to protect their commercial interests within these partnerships and have pondered how the requirements of eBorders have been absorbed into day to day working practices. The presentation of findings has shown how travel firms have adapted to meeting the compliance requirements, but have done so in a manner which allows the commercial interests to be protected. Having examined the strategic level response of the travel sector we now move to look at some unique data from the financial services sector which examines how the customer relationship was implicated there.

CHAPTER SIX

Embedded adaptations

Renegotiating and reworking in the financial services customer relationship

Introduction: Mobilising security in the customer relationship

The connections between the customer relationship and AML/CTF in the financial services sector were unlike that which emerged in the retail travel sector. While the retail travel sector refashioned their eBorders practices so that they explicitly took place within the face-to-face customer interaction, the quality of the financial services customer relationship and how it is managed has been pivotal within AML/CTF from the start. This is because AML/CTF stipulates that participating organisations must not only identify suspicious customers but analyse their ongoing behaviour locally.

Guidelines from the Financial Services Authority Joint Money Laundering Steering Group (JMLSG, 2006) highlight the important role played by customer-facing staff in identifying suspicious activity, the ongoing training and policy requirements as well as the importance of nurturing an AML/CTF culture within the organisation. The guidelines also contain important information about the electronic monitoring of customers, focusing on the areas of greatest risk. Because the regulations have been in place for some time, for the larger banks at least, such operations are now a routine part of customer relationship management activities. However, as Chapter 4 indicated and as the research of Anthony Amicelle has outlined (Amicelle, 2011; Amicelle and Favarel-Garrigues, 2012), it still raises tensions within the organisation that require refashioning. Furthermore, the financial services sector is one of the leaders in the development of customer relationship management practices (Dibb and Meadows, 2004), where customer activity is profiled using data mining techniques so that it can be better understood and so customers can be retained in the long term. Like customer management, anti-money laundering also requires financial institutions to have a good knowledge of the normal and reasonable account activity of their customers, so that they have a means of identifying transactions which fall outside of the regular pattern of activity (Basel 2001, IFAC 2004). Effectively CRM concentrates on the customers that organisations *do* want, where AML/CTF concentrates on those that the organisations *do not* want. Increasingly, organisations use similar technologies to both ends. They use data mining applications to identify risky groups of customers (Levi and Wall, 2004) and criminal association (de Goede, 2012), and to predict future consumption behaviour (McCue, 2006). The question arises as to whether there are synergies between traditional CRM capabilities in the industry that support market monitoring and customer insight on the one hand and those required for AML on the other.

The role of banks in crime reduction

We explore this issue further by drawing on our survey and interview data to examine whether competence in AML is positively influenced by a number of established dimensions of CRM implementation. We also explore whether financial institutions are required to undergo significant changes in data capture, data analysis and staff training due to AML/CTF. We begin by reviewing existing literature on AML/CTF in the financial services sector and on CRM in general.

Although the JMLSG guidelines are extensive, the wording in the regulations focuses on what the financial institutions ought to achieve, but it does not specify how they can or should do it. It is difficult to spot suspicious transaction patterns because they differ from legitimate transactions in very subtle ways (OTA, 1995). Furthermore, money laundering results from a broad variety of underlying criminal activity from tax evasion to drug trafficking, fraud and embezzlement (Yeandle et al., 2005), which can occur together (Canhoto and Backhouse, 2007). Furthermore, the form that money laundering takes evolves, as criminals innovate with new methods (Harvey, 2005), such as moving money over the Internet, or using online casinos and e-banking, quickly rendering obsolete any existing knowledge about money laundering methods. The variety of criminal activity and the evolving approaches mean that there is a low incidence of each particular type of money laundering behaviour compared to the overall volume of financial transactions, making it difficult to develop tested profiles. Allied to this is the ethical problem of false positives: there many technical limitations arising from the lack of reliable profiles, as well as numerous examples of selective and erroneous analysis and interpretation of transaction data. If someone's assets are frozen by mistake, their lives will be very seriously affected (de Goede, 2012). On BBC Radio 4's 'Moneybox', broadcast on 4th January 2014 it was reported that nine citizens of Iran, who were customers of The Royal Bank of Scotland and National Westminster Bank, had complained to the Financial Ombudsman because their accounts had been frozen despite being innocent. The banks argued that mere citizenship of those countries constituted an excessive financial risk and accounts would be automatically closed if it the customers disclosed their citizenship to the bank (BBC, 2014). Similarly Canada's TD Bank closed the accounts of Iranian-Canadians because of Iranian sanctions, even though the citizens were resident in Canada and innocent of any crime (BBC Persian, 2012).

AML practices also run counter to the traditional strategic objectives of and cultures within banks (Donaghy, 2002), as well as the way in which employees' performance is assessed and rewarded (Canhoto, 2008). Furthermore, developing customer insight is a bespoke activity, whereas AML/CTF profiling requires a generalised approach (Kingdon, 2004). Hence, while banks may be well positioned to monitor and prevent the movement of funds associated with criminal activity, AML/CTF presents specific challenges concerning performance management, industry norms and existing customer management practices.

Knowing the customer for relationship management and security ends

In relation to CRM, the first thing to note is that organisations – even those within the same sector – may be at very different stages of evolution in their CRM journey (Karakostas et al., 2005) and consequently CRM is undertaken with varying degrees of sophistication. This ranges from an organisation simply using market segmentation to identify and target attractive customers, to it using full CRM systems that enable organisations to gain a closer understanding of their customers over time (Dibb and Meadows, 2004). There are a number of established dimensions to this journey and it is possible to compare different organisations in terms of their CRM sophistication as well as to use some of these dimensions to explore their AML/CTF capabilities.

The first of these dimensions is the extent to which CRM solutions are seen as a key tool for firms to collect data and develop insight about their customers (Johnson et al., 2012). In all industries who practice customer relationship management (CRM), and in financial services in particular, customer knowledge is enhanced by bringing together customer-related information from multiple sources in the organisation (Sheth, Sisodia et al., 2000) as well as from outside of the firm (Richard, Thirkell et al., 2007). Integrating information from multiple viewpoints allows firms to develop a more detailed understanding of their customers and how they use their products (Sethi, 2000). CRM systems, in particular, allow individual customer behaviours to be monitored (Rollins, Bellenger et al., 2012). Such systems allow firms to identify the lifetime value of the various customers in their portfolio (Lindgreen, Palmer et al., 2006) and support the development of differentiated marketing approaches to the various customers (Zablah, Bellenger et al., 2004).

The second dimension concerns the presence of strategic pre-requisites which define how CRM features in the firm's overall strategy (Bohling et al., 2006). Specifically, CRM activities are shaped by the senior management's belief in the role of relationships with key customers in the creation of shareholder value (Payne and Frow, 2005). Moreover, CRM implementation is influenced by the organisation's culture, namely whether there is a clearly articulated vision regarding the importance of identifying customer needs and whether performance systems explicitly consider CRM activities (Meadows and Dibb, 2012). In our empirical work we looked for evidence of an association between established pre-requisites for competence in CRM and competence in AML.

The other influencing factor in the CRM journey is the extent to which the organisation has an in-depth understanding of the broad range of people, processes, operations and technology issues associated with CRM implementation (Bohling et al., 2006). Finnegan and Currie (2010) argue that CRM success is hindered by an excessive focus on specific software packages. Instead, software deployment should be secondary to defining and developing appropriate customer-facing business processes, including the recruitment and training of customer-focused staff (Johnson et al., 2012).

Dibb and Meadows (2004) review the capabilities that support CRM and classify them on two dimensions, namely hard versus soft factors. The first type of hard factor concerns the technical systems, processes and databases that support the collection and use of customer data. While CRM processes may function with little investment in technology (Keramati, Mehrabi et al., 2010), in today's data-driven environment many firms will deploy information technology systems to support their CRM initiatives (Johnson et al., 2012). We therefore explore the idea that competence in AML is likely to be positively influenced by a number of established dimensions of CRM implementation concerning the strategic deployment of technology.

A tenet of relationship marketing is that the development of relationships between firms and customers is based on commitment and trust (Morgan and Hunt, 1994). Hence, the second type of hard factor identified by Dibb and Meadows (2004) is the capability to understand customers and their needs, calculate acquisition and retention costs and establish a program of regular contact with consumers. Johnson et al. (2012) argue that the main function of CRM technology is to support customer capabilities. We contend that competence in AML may be positively influenced by a number of established dimensions of CRM implementation concerning customer data and knowledge capabilities.

Dibb and Meadows (2004) also identify two types of soft factors. The first concerns the company's implementation of initiatives that support a 'one-to-one future'. This phrase refers to whether the organisation has a long term commitment to understanding customer needs such that they are able to target products and services to the customer on a one-to-one basis. The adoption of distribution channels or the development of products that underpin a close relationship with customers would be examples of this. It has been shown (Grewal et al., 2001) that the success of marketing initiatives is determined by the motivation of the firm to embrace change and the resources it dedicates to implement these initiatives. Programmes to support the development of relationships with customers are not an exception – they require concerted effort across the organisation in order to succeed (Vandermerwe, 2004). A key company capability is the ability to integrate and share information beyond functional units (De Luca and Atuahene-Gima, 2007). We therefore explore the notion that competence in AML is positively influenced by a number of established dimensions of CRM implementation concerning the organisation's strategic focus on customer centricity and a one-to-one future.

The second type of soft factor considers the role that employees play in retaining customers, how they communicate with customers and whether they are empowered to make decisions that support the relationship. The organisation's customer-facing employees are very important in CRM. Employees' affective commitment, for instance, is a key factor in the successful implement of CRM initiatives (Shum, Bove et al., 2008). In addition, employees need to possess the necessary technical skills to be able to use the system to its full potential (Keramati, Mehrabi et al., 2010). Employees that have access to customer insight and are part of cross-functional teams are deemed to be more responsive to customer needs and better able to assess relationship development opportunities (Sethi, 2000; Troy, Hirunyawipada et al., 2008). We now present the results of the survey which examines some of these questions.

Findings

Analysis of survey data

Table 6.1 provides an overview of the survey instrument and the topics covered. The survey was based on previously validated scales (Dibb and

Meadows, 2004; Meadows and Dibb, 2012) which identify and elaborate upon the dimensions of CRM sophistication. These were complemented with questions to explore the extent and impact of AML initiatives; a set of new survey questions on the topic of AML were drawn up by the research team, based on prior knowledge of the relevant literature and of the previously validated scales for CRM that were informing the questionnaire design. Respondents evaluated items on a scale from 1 to 7, with opposing views at either end of the scale.

Section	Topics covered (illustrative)	Number of questions
1A. Responder's profile	Job title; involvement with CRM and/or AML	3
1B. Organisation's profile	Location, turnover, number of staff, areas of banking and finance in which they operate	4
2. Pre-requisites for CRM	Support for CRM from senior management and organisational culture	7
3. Implementing CRM: The Technology	IT as a strategic tool or simply to record transactional data; range of customer data available to staff	9
4. Implementing CRM: The Customers	Focus on value today or on lifetime value of customers; company's ability to anticipate and respond to customer needs	6
5. Implementing CRM: The Company	Focus on individuals or on groups of customers; focus on customers and their life events or on transactional marketing	9
6. Implementing CRM: The People (Staff)	Use of customer contact as a market research opportunity; the setting of appropriate objectives by senior management	6
7. Capturing customer data for AML activities	Possible changes in data capture, data analysis and customer contact due to AML	15

Table 6.1. Structure of the survey instrument.

Findings from the survey data

The respondents represented a broad range of financial services organisations, reflecting the fact that the requirement to collaborate with government in the prevention and detection of money laundering and terrorism financing applies to the whole industry, regardless of the size of the business (FSA, 2003). Around half of the respondents were from large organisations, with turnover exceeding £500m and/or more than 3000 employees. The respondents represented a range of organisations, from building societies (6 %) to insurance companies (21 %) and including high street (17 %), investment (14 %) and online banks (16 %), among others (26 %).

Table 6.2 (below) provides the mean and standard deviation for each question in Sections 2 to 7 of the survey; correlations within each section of the survey are presented in Appendix A.[10] Parts 2–6 of the table (Pre-Requisites for CRM, and the four dimensions of CRM implementation) reveal a range of CRM sophistication levels among the organisations responding. Just over 40 % of respondents reported that their organisation had a CRM team. In terms of the pre-requisites for CRM, most respondents reported that their organisations had a strong stated desire for relationship management and an organisational culture that had a flexible approach to innovation and change. Many respondents felt that CRM had a strong champion in their own organisation and that CRM projects were proactively supported by senior management – but such views were not universally expressed.

The final section of Table 6.2 (Capturing customer data for AML activities) shows that respondents tend to agree strongly with the statements that front-line staff have clear responsibility for reporting on unusual data that they observe, and that staff training has changed as a result of AML. In addition, these projects seem to be strongly supported and permeate all

10 To achieve high levels of scale reliability, the original seven questions in Section 2 of the survey were reduced to four, yielding a Cronbach's alpha value of 0.884. The nine questions in Section 3 were reduced to five, yielding a Cronbach's alpha value of 0.632; the six questions in Section 4 were reduced to three, yielding a Cronbach's alpha value of 0.651; similarly, the nine questions originally in Section 5 were reduced to four, yielding a Cronbach's alpha value of 0.774; and the six questions in Section 6 were reduced to two, yielding a Cronbach's alpha value of 0.636. In addition, the questions concerning the AML requirements (Section 7) were reduced from 15 to 12, as explained in relation to Table 6.3, which follows later.

parts of the organisation. Importantly, the data also suggests that the ways in which data are captured and analysed have changed as a result of AML. Hence, there is support for the notion that AML requires changes in data and staff practices.

Section	Question	Mean	Std-Dev
2. Pre-requisites for CRM	1: There is no stated desire at all within the organisation for relationship management	5.40	1.54
	2: CRM does not have a strong champion at the top of the organisation	4.99	1.79
	3: Senior management is not at all proactive in supporting CRM projects	4.99	1.63
	4: The organisational culture is not well suited to supporting CRM	4.74	1.46
3. Implementing CRM: The Technology	1: Emphasis is on using information to record transactions rather than as a strategic tool	3.84	1.79
	2: Front-line staff have access to only very basic customer data when handling customer enquiries	4.46	1.74
	3: Computer system design and implementation are driven by internal accounting needs rather than external customer needs	3.68	1.60
	4: Systems do not have access to attitudinal/buying behavior data required to identify 'life events'	3.15	1.74
	5: Those handling customer direct marketing never co-ordinate their activities with front-line staff	4.49	1.41
4. Implementing CRM: The Customers	1: Emphasis is on value to be achieved from customers today (perhaps through the sale of an additional product) rather than on customers' life-time value	4.24	1.93
	2: During customer contact the emphasis is always on conducting transactions rather than on updating customer information systems	3.53	1.41
	3: The company is very poor at anticipating and reacting to customer needs (events-based marketing)	4.05	1.76
5. Implementing CRM: The Company	1: Emphasis is on transaction-driven marketing rather than customer-driven/life event-led marketing	4.36	1.57
	2: The company always focuses on customer groups rather than the individual	3.76	1.58
	3: The company focuses on increasing sales volumes rather than relationship building as the route to competitive advantage	4.16	1.86

	4: CRM implementation does not permeate all parts of the organisation	3.55	1.81
6. Implementing CRM: The People (Staff)	1: Senior management never sets objectives which reflect the company stance on CRM	4.24	1.61
	2: Staff never use day-to-day contacts with customers as a market research opportunity	3.60	1.62
7. Capturing customer data for AML activities	1: The ways in which we capture data have changed as a result of AML	2.76	1.85
	2: The ways in which we analyse our data have changed as a result of AML	2.94	1.88
	3: Our approach to dealing with customers has changed as a result of AML	3.12	1.69
	4: The frequency of our customer contact has changed as a result of AML	4.09	1.67
	5: Staff training has changed as a result of AML	2.10	1.39
	6: The data accessible to front-line staff has changed as a result of AML	3.55	1.79
	7: AML has changed the job role of our front-line staff	3.78	1.66
	8: AML has changed the job role for those handling customer data	3.50	1.72
	9: AML projects are supported very strongly	2.64	1.76
	10: AML permeates all parts of the organisation	2.70	1.78
	11: Our organisation has designed systems with AML in mind	3.08	1.64
	12: Front-line staff have clear responsibility for reporting on unusual data they observe	2.07	1.67

Table 6.2. Descriptive statistics.

Turning to the two 'hard' dimensions of CRM implementation – the use of technology and the need to build and use customer data effectively – data shows that the staff in many of the responding organisations still feel constrained by the limited range of customer data that they have available to them (for instance when handling customer enquiries). For example, some front-line staff still don't have access to data on customer attitudes, behaviours and life events. There may also appear to be a lack of coordination in some organisations between staff in front-line roles and those who are more removed from the customer-facing side of the business, for instance organising direct marketing activities. This may explain why respondents from some organisations report that when dealing with customers, their focus is mainly transactional, rather than – for example – updating the customer data that is available on the system.

The survey also explored the two 'softer' dimensions of CRM implementation, around the company itself and the role of staff. Here, it appears that there is less strong support for the notion that CRM projects permeate all parts of the organisation, in comparison with AML activity; support is typically felt to be less strong than for AML projects. Respondents report a tendency in some organisations to focus on customer groups rather than individual customers, and on transactions rather than on customer life events, even during day-to-day contacts with customers that present market research opportunities. The overall picture is one of a wide spectrum of progress with CRM, with some organisations reaching much higher levels of sophistication than others.

We now explore each section of the survey (2 to 6) in turn, in order to understand how AML competence is influenced by each of the established key dimensions of CRM competence. In each section we illustrate the correlations with a Venn diagram for ease of reference. The question items related to AML populate the diagrams. The circles of the Venn diagrams represent the question items related to CRM.

AML and CRM as mutually influential?

Our initial correlation tests strongly suggest that those organisations who are more strongly orientated towards knowing their customers, gathering data about them and managing their relationship with them for commercial ends, will be more geared up towards knowing them for regulatory ends as well. It seems common sense to suggest that those organisations who can identify the customers that they do want will be able to identify better the ones that they do not want, but this is what our data are suggesting. Higher levels of CRM sophistication suggest, therefore, that staff will be better equipped to deal with AML queries by having appropriate information systems and appropriate levels of support for those activities within the organisation.

We found particularly strong and significant clusters of correlations between AML competence and the organisation having the pre-requisities for CRM as well as having deployed technology in a strategic way in order to support CRM. We explore these two clusters in more detail.

AML competence and the established pre-requisites for successful CRM

Statistically significant correlations (as reported in Table B1, Appendix B) were found between four items representing known pre-requisites for CRM (labelled Q1 to Q4) and eight items representing key AML issues (labelled Q5 to Q12). We see strong links between CRM pre-requisites and competence in AML. Visible support for CRM, demonstrated for example via a strong championing of CRM projects and an articulated strategic appetite for successful CRM, is associated with changing job roles around AML, both at the front-line and in data-handling areas of the business. It is also associated with new modes of data analysis and a clear focus on staff responsibilities around CRM (such as reporting on unusual data); it is related to staff training and to changes in the organisation's systems to reflect the demands of AML. Support for CRM is associated with support for AML and with the recognised need for both initiatives to permeate all parts of the company (see Appendix C). Figure 6.1 outlines the relationships in more detail.

AML competence and the strategic deployment of technology for CRM purposes

Statistically significant correlations (as reported in Table B2, Appendix B) were found between four items representing the role of technology in the implementation of CRM (labelled Q1 to Q4) and five items representing key AML issues (labelled Q5 to Q9). These findings point to significant interconnections between areas of CRM competence related to technology, such as improved access to customer data related to CRM initiatives, and a range of areas of progress with AML, such as data capture, data analysis, support for AML projects and the notion that AML concepts are permeating throughout the firm. Systems change to support AML may also be connected to key CRM ideas such progress with the concept of the strategic role of data in the firm and improved data access.

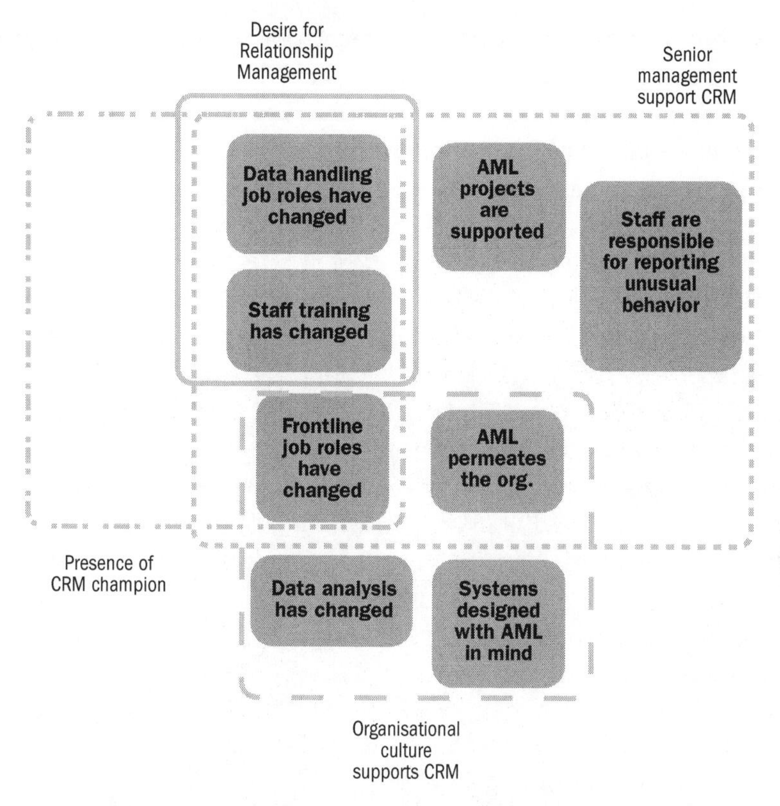

Figure 6.1. CRM pre-requisites and AML correlations.

Once again, these findings indicate associations between positive steps associated with customer knowledge in CRM projects on the one hand – such as the desire and ability to get to know the customer better and to anticipate customer needs – and competence in AML on the other hand – such as making appropriate changes to systems, to data capture and analysis and to staff roles and training (Appendix D). Figure 6.2 (see next page) represents these relationships in graphic form.

AML competence and organisational and staff management dimensions of CRM

Significant correlations were found in three further areas: the relationship between AML competence and customer knowledge capabilities, the organisational dimensions of CRM implementation and between the support and management of staff behaviours around CRM.

We found that where organisations had designed systems with AML in mind, they were also adept at anticipating and reacting to customer needs, and updated customer information systems during the customer interaction. Our analysis also revealed that particular dimensions of CRM implementation relating to the organisation itself provide significant support for AML competence. Competence in key areas of AML – such as making effective changes to job roles and to approaches to customer handling and to driving such changes into all areas of the company – is supported by a number of dimensions of progress related to CRM concerning the organisation's strategic focus. In our data this refers to a paradigm shift from transactional marketing to customer relationship-based marketing which recognises the need to focus on the individual customer and to build relationships over time rather than focusing on the short-term sales of products and services.

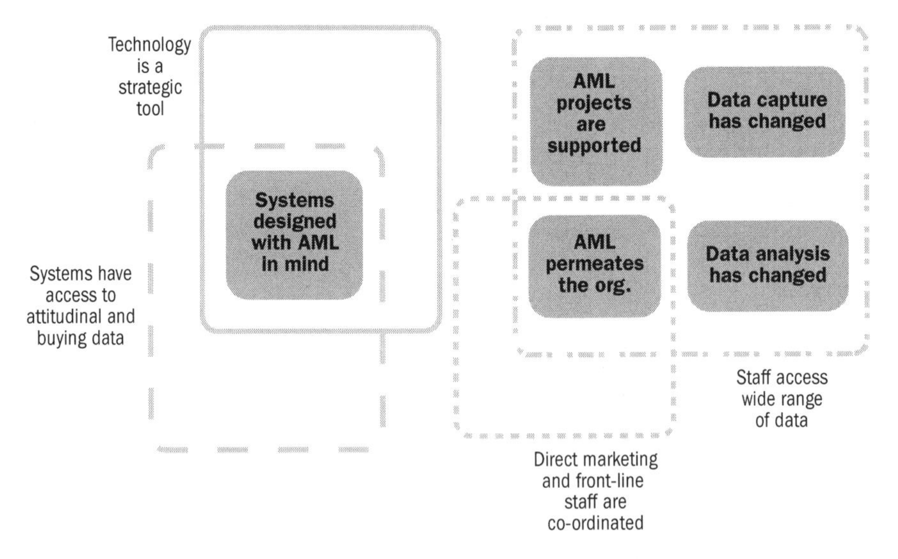

Figure 6.2. Strategic deployment of technology correlations.

Finally we found some significant correlations between AML competence and the dimensions of CRM which concerned support for and behaviours of staff. This emphasises the importance of staff behaviours and management – as supported by positive CRM experience, such as objective setting to promote progress with CRM and using regular customer contact opportunities to get to know the customer better – in moving AML initiatives forward, e.g. making better available to staff, changing the frequency of customer contact and changing staff training to support AML. Graphics and detailed discussion of these correlations can be found in Appendix E.

Factor analysis of the survey data

To take our analysis of the respondents' views further, an exploratory factor analysis (EFA) was conducted. The analysis examined the underlying constructs represented by the measurement items. A factor analysis with Varimax rotation was used to assess the 15 measurement items developed for AML requirements. The results of the EFA after Varimax rotation are presented in Table 6.3. Following prior studies (e.g., Lai, Wong et al. 2010), items with a factor loading of 0.5 or above were retained and items with a factor loading below 0.5 were eliminated to purify the model (Hair et al. 2009). Twelve items remained in four factors, accounting for 75.1 % of the total variance. Construct dimensionality was examined by performing Bartlett's test of sphericity and the Kaiser-Meyer-Olkin (KMO) measure of sampling adequacy. The former was 392.3 (p = 0.000) and the latter 0.768. The values suggest that the intercorrelation matrix contains enough common variance to make factor analysis worth pursuing (Norusis, 2002).

	Factor			
	1 alpha 0.835	2 alpha 0.769	3 alpha 0.912	4 alpha 0.684
1: The ways in which we capture data have changed as a result of AML	.781	.112	.051	.283
2: The ways in which we analyse our data have changed as a result of AML	.880	.074	.108	.133
3: Our approach to dealing with customers has changed as a result of AML	.778	.272	.049	.031
4: The frequency of our customer contact has changed as a result of AML	.569	.484	.204	-.364
5: Staff training has changed as a result of AML	.588	.128	.272	.456

6: The data accessible to front-line staff has changed as a result of AML	.326	.699	.200	.089
7: AML has changed the job role of our front-line staff	.009	.843	.125	.172
8: AML has changed the job role for those handling customer data	.322	.696	-.048	.443
9: AML projects are supported very strongly	.078	.110	.942	.155
10: AML permeates all parts of the organisation	.174	.148	.907	.208
11: Our organisation has designed systems with AML in mind	.048	.259	.272	.731
12: Front-line staff have clear responsibility for reporting on unusual data that they observe	.252	.117	.143	.795

Table 6.3. Results of exploratory factor analysis.

The four resulting factors are labelled as follows: Gathering and Analysing Customer Data, Changing Job Roles for Staff, Organisational Support for AML and Systems and Responsibilities. To test the reliability of each factor, their Cronbach's alpha reliability values were examined; all values were acceptable (Hair et al., 2009; Lai, Wong et al., 2010). It is important to note that this is an important step in the development of reliable scales for constructs supporting AML. While earlier sections of the survey drew upon previously validated scales, the AML section of the survey contained new questions; hence the importance of identifying reliable constructs around AML, as shown by the alpha values in Table 6.3.

The first factor concerns the first five questions in Table 6.3. It focuses on changes in the nature of customer data gathering and analysis – what data is captured and the ways in which it is captured. It also refers to changes in the organisation's approach to dealing with customers including the frequency of customer contact; this is backed up by changes in staff training. The second factor concerns changes in job role, both for front-line staff and those handling customer data. It also draws upon related changes for front-line staff, such as the customer data that is accessible to them. The third factor concerns wider organisational issues such as the support for AML projects, and whether AML permeates the whole organisation. The fourth factor is about the clear responsibility of front-line staff to report on unusual data and the design of systems with AML in mind. Overall, while the four factors are drawn from the AML questions on the survey, they suggest a strong connection with the four dimensions of CRM implementation discussed in the literature (Dibb and Meadows, 2004). The first factor is concerned mainly with issues of Customer Data gathering, the second with Staff, the third with the Company itself and the fourth with Technology. This once again suggests the strength of the inter-connections between practices in the two areas of CRM and AML. These ideas are explored further via the qualitative data that is discussed below.

Analysis of interview data

The quantitative data revealed that four factors – customer data; staff job roles; organisational support and system design and reporting – seemed to underpin survey responses to questions about AML/CTF. We noticed that these factors mirrored those factors which also underpin CRM sophistication discussed in the previous section (Dibb and Meadows, 2004): data and knowledge; strategic support; staff; systems and tools. We then explored the qualitative data in terms of these four issues to reveal the nature of the relationship between the practices of AML/CTF and CRM. It is here that the market logics which justify AML/CTF compliance begin to emerge.

The interviews conducted with key informants and employees of the case study organisations reveal that AML has become routinely embedded into the work of larger financial institutions, which have benefitted from economies of scale and from sophisticated computer systems to help identify suspicious transactions. Smaller firms have found compliance costly by comparison. AML/CTF regulations require the capture of data at great expense to the financial services sector and there seems to be no commercial return on this investment. However, a consistent finding emerging from the case study interviews was that the industry has taken the practice of AML/CTF, examined and addressed the commercial tensions it poses and eventually treated it as a positive aspect of their business. Despite there being no profit in compliance, the practice is deemed to increase the industry's legitimacy and safety. These themes pervaded the discussions that we examined using these four factors as a coding strategy for the qualitative data. As in the case of the retail travel sector the tensions between regulation and commercial priorities were overcome by companies rationalising what was required and incorporating these requirements into the over-riding market-driven logics of customer service, customer retention and commercial survival.

Factor 1: Gathering and analysing customer data

Factor 1 referred to how customer data is captured and used by the organisation and how this is supported by staff training. Within CRM the manner and mode of customer data gathering and knowledge generation about the customer is a key element of CRM practice. In the interview data we found evidence that employees used tacit knowledge about consumer behaviour in general and mobilised it when identifying suspicious behaviour or a suspicious transaction by re-interpreting the account data they had access to. For example:

> And the account has had £200 in, maximum, for four years and all of a sudden you've got £35,000 coming in – well, where has this come from? Obviously it just sends alarm bells ringing but that's the kind of vigilance and awareness you need to at least try and have so that the bank doesn't end up losing money.
>
> (F28, Branch Manager, Company B)

The data also revealed that this knowledge mobilisation only took place at operational, front-line level. Even though there are clear overlaps between the skills for AML and for CRM, there is little cross-fertilisation between the two at a departmental level. Data generated through AML processes were very rarely shared across the organisation (for example, to inform consumer profiles). Where any sharing of information took place, it was done on a 'need to know' basis, as illustrated by this quote from a member of the financial crime team:

> ... The only team that our work sometimes feeds into is we have obviously our internal security and risk management teams. Now they deal with sometimes it could be internal matters, or it could be external matters where maybe we feel a customer is being scammed, so we can then feed that information on to them and say what we think: you need to look at this. But as regards any other department, we don't feed it to anyone.
>
> (F23, Financial Crime Employee, Company A)

Another financial crime employee explains how they may use profiling (and later, contact with the company's marketing department) to identify customers who are likely to be the subject of a particular kind of scam:

> This is a few hundred customers that have been ... subjected to this boiler room fraud; do we have any customers that look like that? So, we'd do a profile on those. The idea behind that is then to be able to potentially ... write out to all the people that look like that to say just be aware you might be vulnerable to boiler room type scams.
>
> (F18, MLRO, Company A)

Interviewees acknowledged that more could be made of warning customers in this way. They observed that this could then be made a feature of the organisation's reputation: one which took the protection of customers' accounts seriously, thus creating a potential source of competitive advantage.

Factor 2: Changing job roles for staff

Factor 2 highlighted how AML/CTF had precipitated some changes in job roles. The interviews with key informants and case study employees explored the intersection between AML and CRM as it was enacted in the customer-facing role, which was initially highlighted by the survey. The intersection between customer service and AML/CTF obligations is demonstrated in the following excerpt:

> If we're not vigilant in the branch, that's going to affect the customer, they're going to give us a bad feedback, they're not going to be happy, so therefore they're not going to recommend us to other people, then have a knock on effect with our sales.
>
> (F28, Branch Manager, Company B)

Some interviewees noted how difficult it was to balance the AML and CRM activities. Two specific activities, informed by CRM, enabled customer-facing staff to carry out their AML activities effectively. One concerns how the front-line employee maintains a normal sales or customer service encounter when they have deep suspicions about the customer:

> ... we used to say, what we call dancing with customers. And I still think it works, because the way it is, you speak to a customer and you build a rapport with them. And it is like you're dancing, because, after a while, you speak to a customer and you know what will work and what won't. If I'm approaching some customers, you've got to be stone cold, you've got to be to the point, and some, you can have a bit of a laugh with them. And you build a bit of a rapport.
>
> (F25, Call Centre Employee, Company A)

Factor 3: Organisational support for AML/CTF

Factor 3 addresses the importance of strategic support for AML projects and whether it permeated the whole organisation. This was a clear issue for interviewees. AML systems and CRM systems are designed with inherently differing intentions; one profiles for risk, the other for attractiveness. As we detailed in the last chapter, the interviewees felt that there was a 'battle' going on between the interests of the organisation to perform a regulatory duty and to provide a competent service to customers. Herein we find the practical synergies between AML and CRM. There is a strong tension felt by organisations in remaining faithful to their business and to their customers, as well as performing compulsory regulatory roles. How organisations perform these roles influences the systems in place. The sophistication of the systems is dependent on the financial resources of organisations. There is therefore a financial imperative which underpins both commercial and regulatory effectiveness: the more resources are available, the better one can do both of these things. As such there is some recognition, as one interviewee suggested, of 'turning a negative into a positive', and exploring the commercial benefits offered by AML/CTF. Comments from the following strategic level key informants argue that some of the insight gained from AML activities will inform their future thinking about the customer. They then go on to suggest that regulation is a long term issue, that is was pointless questioning it, and that their main priority is still customer retention:

> it would be quite easy to let it defeat you, these sort of things, as opposed to just making it as streamlined as possible, doing as much as you can with the information that you're gathering so that you can try to use it in other

> means, because the [organisation's report on their clientele] will form the basis of the segmentation models that we use going forward ...
>
> It's just there's no point in fighting it because it's not ... we internally cannot influence the law in the short-term and in the current climate we're just lucky to be able to keep going, sort of thing, so you just make do. And the more time you waste on questioning and all, then the less time you have to actually address the clients' needs I guess.
>
> (F12, MLRO, Investment Bank)

It appears that synergies between AML and CRM become ever more apparent in the operationalisation and 'streamlining' of systems. The effective working of these systems is important to the commercial viability of organisations, and they utilize past experiences and their resources to maintain and improve commerciality in the face of regulatory requirements, as well as marketing and business requirements.

Factor 4: Systems and responsibilities

Factor four referred to the systems which were designed for AML purposes and the responsibilities that staff had for reporting. As we noted in Chapter 3, in the case organisations we studied the MLROs estimated that only 40 % of the suspicious activity they spotted was detected by transaction monitoring systems: the rest was detected by staff using the bank's account management system and face-to-face interactions with customers. As an industry representative suggests, the cost of implementing an AML system can be substantial and therefore prohibitive to some organisations and advantageous to others:

> It is not a 'level playing field', the costs can be enormous, for instance, the installation of new systems. As there is often a need for a significant system to capture data, there have been a lot of developments over the past number of years and there must be a balance between practicality and pragmatism ... Often changes can be initiated by responses to crisis, trends often influence regulation ... However, there is a constant battle between maintaining the welfare of the firm and that of the consumer.
>
> (F12, MLRO, Investment Bank)

Whilst this may be prohibitively expensive for some smaller organisations, it seems that even the larger ones we studied do not rely entirely on specialist AML systems. They rely on the everyday systems associated with account management and on the vigilance of their customer-facing staff as well as specialist AML/CTF systems. This is the key point of intersection between AML/CTF and CRM: the majority of AML/CTF activities occur in the interactions between staff and customers on a daily basis as the customer relationship is enacted. This is then mediated by existing systems installed for account management purposes. Staff then report any suspicions they have using an AML reporting system which then feeds up to the financial crime specialists. This is then passed on to the National Crime Agency.

AML/CTF and the customer relationship in retail financial services

The data revealed that there are synergies between the capabilities required for the commercial goal on the one hand and the AML/CTF goal on the other. However, addressing the latter requires specific, additional investment from commercial firms and provokes tensions in existing stakeholder relationships as we explored in Chapter 4. Societal goals such as AML/CTF and the prevention of crime and terror are changing the landscape of financial services; such requirements are increasingly incumbent and important to 'reinvented' forms of business (Porter and Kramer, 2011). However, CRM and AML/CTF are not easy bedfellows, as shown by the qualitative data. AML and CRM requirements may be relatively easy to merge as far as 'hard' aspects are concerned, for instance in the collection and analysis of customer data. However, there are clear tensions created for the 'soft' aspects: the staff who are dealing directly with customers. Still, the effectiveness with which staff conduct a customer interaction and their knowledge of 'normal' or 'expected' customer behaviours, which is strongly rooted in CRM, are critical aspects of AML practice. Evidence of cross-fertilisation between AML and CRM at the operational level of the organisation was identified, i.e. for staff in customer-facing roles. Interestingly, no other specific interaction was found at other organisational levels, e.g. for staff in more 'back office' roles such as those handling customer data; which indicates that the front-line is where AML, like CRM, is really enacted.

The qualitative data explored the challenges around contributing to the government's AML/CTF programme. There were conflicts between the commercial and the AML/CTF goal, particularly at the level of the banking principles of secrecy and privacy, as well as regarding concerns with the customer experience. While the former had been mentioned in the literature (Donaghy, 2002), the latter had not been recognised. It was also clear that the implementation of AML systems and procedures is a very expensive and complex exercise for financial institutions, with significant economies of scale. Moreover, the AML exercise does not contribute positively to profits in the short-term. Despite the significant direct and indirect costs for the commercial organisations of supporting AML/CTF, the organisations could still see the long-term benefits of 'being companies that care' (Kotler, 2012). Specific benefits mentioned included increasing public trust in the system, improving the institutions' image in the aftermath of the sub-prime crisis that is deemed to have triggered the current economic recession in Western economies and acknowledging the role of financial institutions as enablers of the movement of funds worldwide.

Conclusion: Market logics, commercial priorities and the customer relationship

eBorders and AML/CTF are both regulatory frameworks with which commercial firms must comply and which involve partnerships between public and private organisations to ensure security information flows back to government. We will now revisit the two questions we posed at the start of Chapters 5 and 6:

- What actions are taken to protect the commercial interests of firms within these public-private partnerships?
- How are the requirements placed upon these firms absorbed into their day-to-day working practices?

In exploring how firms in the travel and financial services sectors have absorbed the requirements placed on them by UKBA and the NCA into their working practices, it is necessary to understand the implications of each regulatory scheme for firms. eBorders, for example, has a number of distinctive features which have influenced how travel sector firms have responded. First, all carriers operating into and out of the UK are subject to the same compliance requirements; second, compliance is compulsory and enforced; and third, firms must comply with the timeframe laid down by UKBA and the processes that have been established. Consequently, the opportunity for some operators to benefit at the costs of others is limited. Even so, we found evidence that smaller carriers and firms further down the supply chain felt disadvantaged compared with the larger organisations who were seen as having greater power to influence how eBorders was influenced.

Meanwhile our data on the financial services sector supports previous findings that there may be different levels of sophistication within a particular industry, in terms of customer management capabilities, for example (Dibb and Meadows, 2004; Karakostas et al., 2005). In other words, not all players are starting from the same position when required to respond to AML/CTF initiatives. Our results take the previous findings from the literature further by showing how the levels of sophistication can be broken down into a number of elements. Specifically, it was shown that financial organisations were more likely to have high 'hard' CRM capabilities in place than 'soft' ones. Yet, it was the soft CRM capabilities that were more strongly linked with the ability to comply with AML/CTF regulations.

Protecting commercial interests

In both the travel and financial services cases we found evidence that the compliance requirements impacted upon commercial interests. Market logics emerged in both sectors which rationalised regulatory compliance within commercial priorities. For example, travel firms dealing with the eBorders regulations faced major difficulties as a result of the costs and disruption caused by the required operational changes, their perceived need to protect their customer relationships and from negative effects on staff working patterns and morale. The jobs of front-line customer service staff in travel firms' call centres, for example, dramatically changed as a result of eBorders and the escalation of compliance-related customer contacts. Amidst the resulting disruption, a particular concern for firms was how this might affect customer relationships and the ownership of those relationships in the longer term.

In the financial services sector, we found synergies between the technical capabilities of customer insight that support marketing activity and those required for AML/CTF. Nevertheless, there seemed to be barriers in terms of incorporating the skills and insight derived from AML into commercial activities, and this limits the organisation's ability to respond to environmental changes (Homburg et al., 2007). The implications from our studies are that while firms can seek to protect their own commercial interests while working on public-private partnerships and be cognisant of the risks involved in dealing with financial services customers, any commercial benefits are unclear. Industry players appear to experience difficulties in sharing learning acquired from one activity (such as AML/CTF) and applying it to other activities (such as CRM).

Our work in the financial services sector indicates that there is an uneven impact on the various players in the industry, for instance in terms of the costs of the solutions adopted. Specifically, there was a direct relationship between the availability of financial resources and the strategic approach to customer management on the one hand, and the ability to meet the broader goals of the AML/CTF initiatives on the other. While this finding confirms the literature regarding capabilities and performance (e.g. Krasnikov and Jayachandran, 2008; Theodosiou, Kehagias et al., 2012) it also raises important questions regarding the ability of different types of organisations to meet societal goals. This may have implications in terms of the competitive structure of specific industries; for instance, some smaller organisations may be at a disadvantage in terms of both bearing the costs and achieving any benefits from participation in such an initiative.

In the case of eBorders, carriers and other travel firms struggled at the outset to respond to and minimise the disruptive effects of compliance. They initially engaged in 'fire-fighting', doing what they needed to in the short term to comply while at the same time safeguarding their customer relationships (Smart and Vertinsky, 1984). The stakeholders soon became bolder, moving to restore their commercial interests and to secure their competitive positions in the market. The individual responses of firms differed, influenced by issues such as their size, available resources and capabilities – and this is also true in the case of AML/CTF, where the passage of time has allowed a more developed picture to emerge.

Capturing and converting consumer data: Absorption into everyday commercial practice

Our results highlight the synergies and tensions that can occur when private sector organisations work alongside government and re-align resources to develop integrated approaches to solving problems. We have focused on how industry players in both sectors maintained their commercial priorities whilst capturing and converting consumer data for national security surveillance. We discussed some of the ways in which they responded to new external requirements and struggled to balance competing pressures. As well as shedding light on the role of financial institutions in preventing crimes, our results from the financial services industry have shown strong inter-connections between customer management capabilities and money laundering prevention. Our analysis of the travel sector, meanwhile, has revealed how firms become more adept over time at working with the regulatory requirements to restore their commercial interests.

In the financial services sector, we have noted the obvious connections between CRM and AML and the apparent synergies between the two activities (such as the drive for better customer data). However, it is also evident that CRM and AML have contrasting rationales. CRM focuses on building relationships and lifetime value for the financial services organisation, while AML focuses on crime reduction via customer monitoring. These two rationales can 'clash' at the point of the customer interface, but produced the notion of the 'compliant sale' being valuable, lasting and important. This may negatively impact on the customer experience because of the conflicting demands on the employees who are at that interface with the customer, but when executed well, it is a source of added value. This suggests that, looking to the future, managers need to investigate ways of softening this impact, for instance in terms of service design. Some organisations expressed grave concern that the customer service experience could be damaged, on occasion, by AML/CTF activities and therefore financial services firms are likely to seek to minimise the impact of this on their future business interests.

Given the role of front-line staff in service delivery and customer satisfaction, it is likely that the need to reconcile the clashes between the CRM and AML requirements should be addressed at the level of front-line staff roles. Again, managers in the financial services sector need to consider this requirement in terms of staff recruitment and training. We also find evidence of work intensification in both sectors for front-line staff, with more responsibilities, more systems to interact with, more data to handle and, in the case of financial services, the introduction of 'suspicion' into front-line roles; i.e. clear responsibility for reporting 'unusual' activity by customers. This should be recognised in terms of the support systems made available to customer-facing staff, as well as during performance assessment. It is significant that the commercial response in both sectors was to rework how the customer relationship was approached in the light of the regulation. This has, intentionally or otherwise, resulted in a responsibilisation of front-line staff for national security matters in a way which compromises their existing roles in the longer term.

It is too simplistic to approach the discussion of the role of business in society as an either/or situation, as is common in the criticism levelled at business in general (Kang and James, 2007), and marketing in particular. Clearly, there may be conflicts between commercial goals and societal ones, particularly where businesses face the financial cost of solving a public problem. However, businesses that think strategically about this problem can recognise the long-term benefits of addressing societal problems. In the case of AML, the benefits identified were mostly in terms of increased legitimacy and safety of the industry, though more generally benefits may include the fostering of innovation, an increase in productivity in the supply chain and the improvement of relationships with other organisations in the community (Porter and Kramer, 2011; Kotler, 2012). Similarly, it is clear that travel firms have a vested interest in ensuring safety for travellers, yet as they became more used to the implementation process for eBorders they became more positively disposed towards maximising the additional contacts with customers for cross-selling purposes. This is an encouraging finding, as the trend towards involving businesses in solving societal problems is increasing and broadening in scope (Hutter and Jones, 2007). However, the government's use of the private sector to deliver an integrated solution to a societal problem is not without challenges. Both the travel and financial services case studies demonstrate that the nature of the interactions between staff and customers, required to maintain good service relationships, can be in conflict with the interactions required for eBorders or AML. As Braithwaite (2008) argues, it is also possible for firms to do well at both activities, but only if they are already in a powerful market position, either in terms of size, resources, capability or because of their relative command of their supply chains. Furthermore, while the positive impact of employee satisfaction on customer service has been well established, there is a gap in terms of research focusing on employee experiences in the context of the delivery of societal goods (Smith, 2012). In the next two chapters we specifically examine the impact of these two regimes – and the emergence of the market logics wherein they are executed – on employees.

CHAPTER SEVEN

Cross-selling for security

Remediation work at the retail travel front-line

In Chapters 7 and 8 we explore the impact of eBorders and AML/CTF on frontline staff. In Chapters 5 and 6, we outlined how firms involved in both of these regimes rationalised their regulatory obligations in similar ways. In retail travel, after trying to keep eBorders requirements away from commercial processes, airlines, tour operators and travel agents each realised that it was important to incorporate it into the customer relationship. The point of customer contact through eBorders became the opportunity not only to capture passport information, but to improve the customer experience and try to cross-sell optional extras. Travel agents in particular saw passport data collection as an added extra in itself that they could offer to customers to prevent them from being tempted by tour operators or airline websites if they were to enter passport data themselves. In retail financial services, despite the obvious financial burdens imposed by the AML/CTF regulations, firms eventually saw it as an opportunity to cement corporate reputations and to reassure customers that they would be protected from fraud. At the same time, the main operational burden of identifying suspicious activity rested with the front-line customer-facing staff. This was not only because the regulations themselves stipulated staff responsibility for AML/CTF but because, in very practical terms, the point of customer contact with the bank, be that through electronic or face to face means, was where the majority of questions were raised about suspicious customer behaviour. Although transaction monitoring systems picked up some questionable activities and financial crime departments also had a role to play, by the banks' own admission, front-line staff made a large contribution. Therefore, in both retail travel and in retail financial services, a direct consequence of the impact of the regulations on the firms was for them to focus on the customer relationship. First as the primary source of data collection for the regulations, but most importantly at the moment where the regulations become subordinate to the market-driven logics of the private sector firm: where they could re-capture authority and legitimacy and resolve some of the more macro level tensions that we explored in Chapter 4. Frontline customer-facing staff were working at that interface between the regulations, organisational resource allocation and the market. Those who faced customers became the bearers of the regulatory burden on a day to day basis.

Towards remediation work

In this chapter, we develop a new concept 'remediation work' which exposes the additional tasks and pressures to which these staff were subject as a consequence of how their organisations had strategically adapted to the regulations. We use Bolter and Grusin's concept of 'remediation' and extend it to develop the notion of 'remediation work'. Remediation is a term from new media theory which refers to how one medium becomes re-presented in another medium and how the world around that re-presentation becomes reconfigured as a result. Although the term was first coined and explored by Bolter and Grusin (2000) in a more recent incarnation of the term – 'premediation' (Grusin, 2004) – has found its way into discussions of security (de Geode, 2008). This term has been used to explain how the media and cultural industries have already mapped and depicted a whole range of security futures which are subsequently influencing the policy and public imagination. However, in this chapter, we return to the original notion – remediation – to develop the concept 'remediation work' (see also Ball et al., 2014) to describe those areas of consumer behaviour which are not 'premediated' and need to be remediated by someone or something. Remediation work is used to describe the types of activity required of front-line workers by the two regimes and the impact that it has on their working lives. Both of the regimes rely upon the remediation of information about the consumer and their activities in order to function. In financial services, customer activities and their financial transactions become remediated into information infrastructures which support the identification of criminals and terrorists. In travel, customer passports are remediated into information infrastructures which transmit their details back to the UKBA to aid their endeavours to identify the same. Remediation work thus refers to the extra tasks completed by front-line staff, which ensures that information about customers becomes part of the securitised information flows that form the heart of these two schemes. The form that remediation work takes also becomes shaped by the same infrastructures and technologies which surround it. We define remediation work as:

> activities undertaken by front-line customer service workers in a pan governmental-private sector surveillance regime which allow surveillance data about customers to flow back to government in newly-mobile forms.

Remediation work has a number of different elements, the precursors to which have been discussed by other authors in relation to security although not using the same terminology. Amicelle (2011) highlights the misapprehensions which occur between financial services and anti-money laundering professionals at higher levels in the organisation and their divergent interests and objectives. Front-line workers, however, have to reconcile these differences in the moments of the customer interaction. It is the actions and activities undertaken by those employees to *reconcile* the differences between sales-service (Tyler and Blader, 2005) on the one hand and security on the other, which constitutes remediation-work. Front-line workers use techniques from their repertoire of sales–service skills to transform elements of the customer interaction into pieces of information that can then be used for national security purposes. We reveal how they draw upon the performative aspects of interactive sales–service work and their existing knowledge of customer behaviour to pre-empt any security concerns to bridge the gap between sales and security.

It is perhaps not surprising that these are the resources that front-line workers draw upon to discharge regulatory obligations. Performing the sales pitch, reading from the company script or capturing customers' information are part of the routine tasks performed by a frontline customer-facing worker when dealing with customers, colleagues, managers or superiors. Such embodied executions have not gone unnoticed in the wide ranging literature reviewing workplace practices (e.g. Geiger and Turley, 2005). As these works have noted there is value in performing 'good service' designed to enhance commercial advantage. Nevertheless, due to the ephemeral, expressive and emotive nature of these types of performance it has intrigued commentators due to its ambiguous and amorphous qualities, alongside its centrality to business relations and communications. Indeed, performance as a metaphor for the interactions between customer and employee is a widely used theme in many literatures (see Weaver, 2005). From flirting and wooing customers (Bigus, 1972), to the cultures of haggling (Herrmann, 2003) to the institutional practices of mis-selling (Ericson and Doyle, 2006), all help to elaborate the centrality of social interaction to the success of customer-facing practices. The 'fluidity and adaptability of human activity' (Weaver, 2005) is a performance of great influence to the workplace; even when, as Weaver is keen to point out, in heavily structured and scripted environs – the call centre is an example par excellence, where employees follow rigid scripts – even then desired outcomes are not always met. And while work by Goffman (1957) looks to the favourable presentations made to strangers, Crang (1994) extends this further when examining a themed restaurant chain where employees are expected to look and act like particular 'celebrities' and Tyler (2009) examines similar phenomena in children's culture industries. In these instances, performances are tailored to the social encounter in which they find themselves, referred to by Bryman (2004) as 'performance labour'. Indeed the way workers look, sound and act, are themselves part of a service product (Leidner, 1999). Furthermore such performances allow the front-line worker to accumulate embodied and tacit knowledge about how to handle face to face service encounters. Some technical or product knowledge is required, but typically a base of tacit social competences is built, which enable workers to perform the social interactions which characterise face to face service work. Harnessing informal or tacit worker knowledge has long been the subject of management theorising, from discussions about emotional labour (Hochschild, 1983) to invisible infrastructure work (Star, 1991) and from Taylor to Quality Circles and the knowledge economy (Thompson et al., 2001). Throughout these discussions, however, the embedded and embodied knowledge of the routine service worker have been marginalised. By contrast, in this chapter we extend the analysis made in chapter 6 to show how the lived bodies of the travel and financial services front-line workers, in terms of their performativity and embodied knowledge, are pivotal points in both schemes. Indeed the extra work undertaken by employees and the deployment of these under-acknowledged skills, serves to politicise the regimes from a labour perspective. Before going into the empirical detail, however, we first introduce our key concept for this chapter, remediation work.

Remediation work

Re-mediation refers to the incorporation of one medium into another and was originally employed to analyse the significance of new forms of digital media. Bolter and Grusin (2000: 183) state:

> As a digital network, cyberspace remediates the electric communications networks of the past 150 years, the telegraph and the telephone; as virtual reality, it remediates the visual space of painting, film, and television; and as social space, it remediates such historical places as cities and parks and such 'nonplaces' as theme parks and shopping malls ... cyberspace refashions and extends earlier media, which are themselves embedded in material and social environments.

When considering re-mediation one might initially think of popular examples such as the change in the format of commercially available music from CD to MP3 and its impact on the recording industry. But re-mediation is also at the heart of our two surveillance regimes. AML/CTF is premised on the remediation of suspicious customer behaviours and transaction patterns. Once a staff member has identified suspicious activity, information about it is inscribed into the bank's fraud reporting system which is then investigated by its financial crime department. If the activity is deemed sufficiently serious, financial crime employees then submit a Suspicious Activity Report to the National Crime Agency via their web portal. Similarly eBorders is premised on the digital capture of paper passports – in other words, their representation in a new medium – at or around the point of sale of the travel product and before travellers arrive at the airport. The UKBA views eBorders as an improvement on older methods of collecting passport information because it enables advance screening of the travelling population against international watch lists and behavioural indicators of risk (MacLeod and McLindin, 2011; Vakalis et al., 2011). Passports have always been scanned in the airport but their upstream capture in the commercial setting is a new development. Effectively they are re-mediated in a different time and space and by different social actors.

New activities arise from the collection and transfer of data, which connect local working conditions, politics and meaning systems within the organisation to the governmental and organisational information infrastructures that bring the regime into being. These activities construct the 'securitised information flow' from the private sector to government, and activities of remediation – the inscription of new information about suspicious activities into reporting systems – are at its heart. This inscription process is embedded within employee decisions, their main task activities and the organisational resources which are configured to support the practice. By ensuring that data are captured and transferred, front-line workers are the link between the customer interacting with his/her organisation *as the consumer of a financial services or travel product* and the customer becoming inscribed into a government surveillance regime *as a potential threat to national security* (see also Amicelle and Favarel-Garrigues, 2012). Furthermore an important double meaning of remediation work extends to activities of repair (Graham and Thrift, 2007) and some difficulty is expected when combining the information infrastructures of different organisations and the state. The data highlight that compliance with the regimes at an operational level is a highly contingent and emergent activity which bridges gaps and smoothes over the joins between infrastructures, customers, organisations and the government. It is primarily employee activities which achieve this 'bridging' between these elements and it is these activities that we characterise as 'remediation work'. Whilst we use the concept of remediation work to foreground the work demanded of frontline staff by the regimes we also use it to politicise them. Re-mediation theory states that the act of re-mediation refashions the networks of actors, resources and other media that produce it while simultaneously bringing them together (see also Bowker and Star, 1999). Thus, the newly remediated suspicious financial activity and the newly remediated passport have the potential to reconfigure existing social, political and material orders in the contexts that bring it about, including those that affect employees. This is political in the sense that extant political interests tend to coalesce in existing resource configurations within organisations. Any refashioning of these existing configurations will be politically disruptive by necessity, particularly as it would seek to embed a new interest, that of the public interest of security, within private sector resources. As a concept it captures the impact of the regimes on front-line workers and, according to the definition, has two components. The first concerns the act of *transforming media*, i.e. the generation of data from other sources concerning the customers' identity and their activities and its capture into an information system. The second concerns the activities which *reconfigure resources* to constitute a working infrastructure which enables this to happen: as Bolter and Grusin (2000) argue, a refashioning of reality accompanies re-mediation. In focusing on the new activities required of front-line workers, we acknowledge that the information systems and infrastructures which surround front-line workers also exert an influence on how remediation work is carried out. This is particularly the case in the travel industry as we shall demonstrate that the work intensification that took place occurred partly as a result of shortcomings in new system design and with legacy systems not being sufficiently flexible enough. Similarly, the exacting requirements of identification systems in financial services resulted in front-line workers developing a whole new vein of tacit knowledge to ensure that the customer's identity satisfied the systems' requirements for a positive and low risk customer identification. We also acknowledge that through certain analytical lenses, these technological elements would have agency and we would not underemphasise that they have a significant role to play. However, we are keen to focus our analysis on the issues faced by front-line workers and so we will focus primarily on their experiences.

Remediation work in the UK travel sector

Remediation work in the UK travel sector as a result of eBorders highlights how employees enable information to flow through organisational infrastructures to government. However the way they describe it highlights the pressure exerted by the market logics of customer retention and cross selling which dominated the industry's response as a whole. Part of this is because of the immaturity of the scheme as well as its design, but some fundamental elements of the retail travel industry infrastructures made eBorders particularly burdensome for front-line retail travel workers. Retail travel focuses on travel for leisure purposes and the main product is the package holiday. These products, sold by tour operators and travel agents, combine several different elements, such as flights, accommodation, car hire, excursions, insurance and other extras, and can be bought in high street travel agencies or online. The airlines involved are 'charter airlines' and they are just one element of a complex travel product. Yet it is airlines that must transfer passport information to the UKBA. As tour operators charter seats on different airlines, and as they have direct contact with the customer, it has become the responsibility of tour operators and travel agents to collect passport information on the airlines' behalf. Without the help of Global Distribution Systems the retail travel sector had to design their own ways to capture and transfer data under eBorders. It was the nascent nature of these systems that made breakdowns likely because eBorders had effectively re-constituted the point of customer contact as the point of passport data capture. The data that we draw on in this section were gathered in three different organisational sites. Company 1 was a customer services call centre of a large tour operator which also had a strong online presence and its own airline; Company 2 was a large retail travel agency operating on the high street but also with online outlets and Company 3 was a small tour operation business and a small high street travel agency with only one high street branch. In a similar manner to financial services, the data highlight that eBorders compliance at an operational level is an emergent activity which bridges the gaps between infrastructures, customers, organisations and the government. Re-mediation work in travel is now explored empirically.

Transforming media

The primary impact of eBorders was that employees had to ensure that passport numbers and names were transformed into digital information which could be sent to the relevant airline and on to the UKBA. Because remediation work is a different activity to that of their primary employment, its introduction represented a de facto work intensification. It comprised extra tasks for which workers were not rewarded, either formally or informally. In Company 1, problems arose when the customer was asked to enter their passport data into a dedicated eBorders website. The website asked for a booking reference, customer names and the passport numbers of each customer. Customers entered the wrong booking references, they typed their names in the wrong format and when their input was rejected they panicked and called the call centre, where the front-line workers began to re-mediate. Workers took passport details from customers and completed the inputting while they were on the phone. Call centre workers were initially told that the call centre could expect 100 eBorders related calls a week. However, they received 500–600 calls per day, most of which required them either to input information on the customer's behalf or to advise customers how to do so themselves.

From the outset, Companies 2 and 3 took the responsibility of ensuring that the customer had provided their information in time. If they failed to do so, there was a danger the customer would go to a tour operator's website to input their data and get a better deal, losing the travel agency money. Inputting passport data added, on average, 15 minutes to every customer interaction. A branch manager in Company 2 noted 'I moan at them when they're not hitting their targets, and they're going "well, I've got this pile of admin to get through"' (T13, Tour Operator Call Centre Team Leader, Company C). In Company 3, employees commented that eBorders had doubled their workload because of the amount of time spent chasing customers for their passport information and checking that it had been received and inputted in time. The following Company 2 employee explained:

> we were adding a whole operational process into how we manage customers, and indeed a whole operational cost associated with that, because we would effectively go onto the website as the customer, load their details on

> their behalf, meet the obligations that the tour operator needs, but actually have to go through a whole piece of work to make sure that we did it.
>
> (T21, Commercial Manager, Company D)

Reconfiguring resources

The second element of re-mediation work concerns how employees reconfigured the available resources to embed eBorders requirements into their working practices. Employees used existing technologies, procedures, materials and personal insights to ensure passport data were transferred. Three separate activities emerged through which this was achieved: integrating, re-assembling and pre-empting. The data illustrate that resource reconfiguration was an uncomfortable process.

Integrating

Unlike employees in the financial services sector, who were working in a mature scheme, the first challenge front-line workers faced was to integrate re-mediation work into everyday work procedures from a standing start. The interviews highlighted how eBorders was almost a 'dance' where responsibility was passed between customers, employees and their supervisors. Employees and supervisors struggled to prioritise the performative requirements of their roles and commercial responsibilities with eBorders. The Integrating task illustrates how passport re-mediation begins to remediate the travel purchase process itself.

Company 1's experience shows that eBorders requirements sat uncomfortably alongside existing infrastructures and working practices. Employees' lack of control over call volumes, systems and customer sentiments resulted in demoralisation in the face of high numbers of queries. As one supervisor commented, 'If you are taking 80 of the same thing, of course you're going to be demoralised'. To this end, managers at each site encouraged staff to treat eBorders contacts as sales opportunities and to see something which was ultimately dull in a more positive light. Supervisors in site one told the call centre staff to ask frustrated customers about their forthcoming holidays and try to sell them extras, e.g. extra-legroom seats or car hire. This supervisor explained:

> I think it's a sales opportunity for them, and I try and tell them in their call coaching sessions to sell.... we should be appreciative that we've got calls, because we've got a job.
>
> (T13, Tour Operator Call Centre Team Leader, Company C)

However, because the calls were from frustrated customers and because there was sometimes little a worker could do to alleviate for example, a system problem that was preventing the input of data, more tension and frustration resulted. A Company 1 employee described how difficult these situations sometimes were:

> All I know is that the passenger's shouting because they can't travel on Saturday because 'I can't put the details in, and I'm going to cancel that flight' ... I didn't put that on there and I'm sure you can travel ... I understand you're shouting at me because you want to do it, but ... [the booking files] haven't downloaded into [the reservation system] ...
>
> (T12, Tour Operator Call Centre Employee, Company C)

Here the employee refers to systems over which the employees have no control. The booking file took 24 hours to download into the reservations database which powered the eBorders customer website. If customers attempted to enter passport information in that 24-hour period after booking their travel, the company's eBorders website did not recognise their booking. Once the employee had explained this to the customer, there was little willingness on either side to explore sales opportunities.

Staff in Company 3 saw eBorders as a direct threat to their business and so responded by collecting unnecessary information from customers. They were keen to integrate eBorders requirements as quickly as possible in order that they did not lose customers to competitors. The following comment of the agency's general manager highlights the threat of eBorders to the business and how it has intensified the collection of passenger information:

> I think we got to the stage where it was the more information you gather and put it on, then you can't go wrong with it because you don't want to be the one with the person without the information being refused travel.
>
> (T16, General Manager, Company E)

Although different integration activities were observed in Companies 1 and 3, it is significant that one's future in the travel industry, either as an employee or as an organisation, was invoked as a rationale for compliance in both instances. In Company 1, the supervisor stated that one should be grateful for a job, and in Company 3 the general manager strongly implied long term detriment to the business if a customer was refused travel. There was little room for manoeuvre.

Reassembling

To ensure passport data were transferred, existing and new resources were reassembled, effectively bridging the gaps between business and eBorders infrastructures. This was particularly the case in Companies 2 and 3, the travel agencies. For them, eBorders turned into a juggling act between the systems of different tour operators and airlines. Staff described this in detail, demonstrating their deep working knowledge of numerous operator sites and systems so that they could enter passport details. They sometimes had to 'pose as the customer', inputting details as if they were the customer, when the customer was unable to do it themselves. As with financial services, a performative veneer emerged, as the employee performed in the gap between the customer and the information system, but this time with the full knowledge of all involved.

The observation of Company 2's eBorders temp highlights this aspect of re-mediation work. Sitting at a single desk in a large open-plan space she had devised her own way around existing corporate information systems, external websites, self-styled paper filing systems and tallies to manage a baffling array of eBorders requirements. Extraordinarily calm, well organised and diligent she confidently fielded phone calls from panicking travel agents, home workers and customers who were struggling to enter their data, reeling off the requirements of different airlines by heart. The budget airlines accepted eBorders data on booking and would refuse travel to those who had not provided passport detail in advance; some tour operators only accepted this detail within eight weeks of departure; others accepted it up to six months before the departure date; for one airline, a specific reference was needed; for another, if a passport number was not provided, the customer could not book a ticket at all. Passport information also had to be provided for each name on the booking, not just the name of the person who made the booking (known as the 'lead name'). The entirety of this temp's work involved inputting the passport information of different customers into different systems at the right time. The complexity of the work she undertook was typical of the way in which the everyday jobs of every retail travel agent had changed as a result of eBorders.

Back office organisation had changed too. Staff in Company 3 had restructured their entire filing system to ensure that the customer had provided their passport information. As they used paper files, they devised checklists to keep track of customers' passport details. Each retail travel interviewee reported how their desks had become piled high with paper files because of the number of outstanding passport queries:

> We have in our files, I can show you paperwork, in our files we would be thinking 'right passenger's booking form you're required to do that' and then we need to check that they've physically done it and write 'yes passengers have done it on such a such date' to keep ourselves right. So it's all about I suppose covering yourself.
>
> (T16, General Manager, Company E)

The interviewee's use of the phrase 'covering yourself' indicates the increased levels of responsibility and even liability felt as a result of eBorders.

Pre-empting

The final activity, pre-empting, involves the employees' deployment of their tacit knowledge of travel customer behaviour to second guess the customers' propensity to either ignore or misunderstand eBorders requirements which were typically described in the 'small print' of their bookings. Customers missed the eBorders requirement in different ways: by misreading obvious instructions, by being too laid back about the importance of passport information or by not being an experienced enough traveller to know what to look for. In describing this phenomenon interviewees in Companies 1 and 3 also acknowledge that this is a source of extra work. In Company 3, sales staff were aware that customers were likely to ignore the small print and pursued customers relentlessly for their information:

> We would need to be chasing them to make sure [they'd] done it because nine times out of ten a lot of people don't read all the small print and with the confirmation you probably get about six pages of it. They probably look at the front bit and go, yes great gone.
>
> (T16, General Manager, Company E)

Imploring passengers verbally was another way travel agents handled this duty. Staff at Company 3, already annoyed by the burden of eBorders, were frustrated by the attitude of some passengers, given the extra work required:

> It's just more time and no money for it. And the annoying thing is with some passengers that maybe don't travel very often they're just like 'oh it's okay I'll bring the passport in one day, it's nothing to worry about'. But you can't stress enough with some destinations, it has to be in, you've got to come in with it.
>
> (T19, Retail Manager, Company D)

Workers in Company 1 also reported that experienced travellers would enter their data correctly and follow the right instructions at the right time. Others, who one worker complained 'didn't use their head', would phone them in a panic, prompted by the company website:

> It's quite clear enough for somebody that travels a lot and uses their head, but there's a lot of people that it's not clear for; even though it's there in black and white – 'please do this' … It needs to be bigger and it needs to be bolder. It needs to be really obvious, but I think the boldest bit is our phone number.
>
> (T19, Retail Manager, Company D)

Whether passengers were unthinking, laid back, evasive or reluctant to provide their passport data, these scenarios demonstrate the lengths to which front-line workers went, as well as the frustration, anxiety and stress they reported experiencing, to ensure passport data were captured and transferred in time. As well as having to input data, employees worked to bring the customer, with their passport, to the right information system at the right time, while not damaging the interests of their employer or impacting their own outputs. These activities, as captured through the concept of re-mediation work and its associated activities, describe the operational burden of the programme and some of its hidden costs.

Conclusion

As the data illustrate the eBorders programme has placed increased workload and additional responsibilities on front-line staff in a tour operator and two travel agents in the retail travel sector. This work is done for no extra reward, it is not integrated into the criteria against which employees perform and so is perceived as an unwelcome regulatory burden. Thus, the activities of integrating eBorders into business processes (integrating), bridging gaps between business and regulatory infrastructures (reassembling) and employing tacit knowledge of customers (pre-empting) arose, representing an intensification of work. This was particularly the case in Company 1 (a tour operator), in which the workload was directly affected by a badly designed customer website, which has since been rectified. However, in Companies 2 and 3 (retail travel agencies), the nature of front-line work had permanently been altered. Agents were required to gather and input data into a wide array of tour operator websites under the threat of losing the customer, revenues and, by extension, their jobs. Other infrastructural elements include not only customers, websites, call centres and companies but also national and international legal systems, industry practices, industry structures and systems, each of which has a constitutive effect on the programme as it was experienced by front-line workers. In the next chapter we explore how remediation work played out in the Financial Services sector.

CHAPTER EIGHT

Compliance conquers all?

Remediation work at the financial services front-line

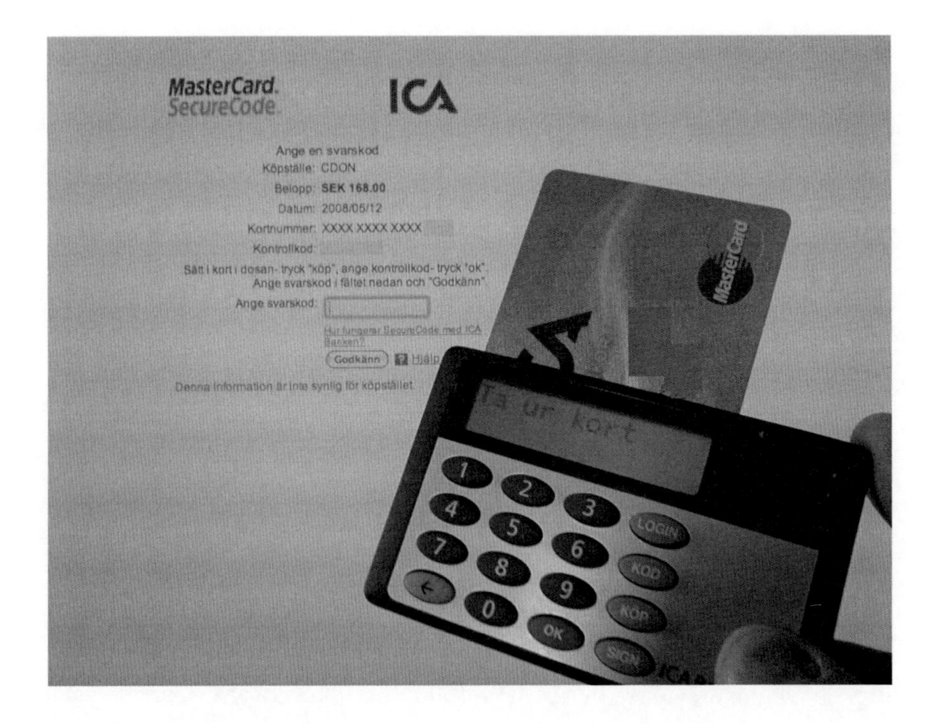

The data that we draw on in this section were the case study data collected in a retail bank (Company A) and a building society (Company B) in the United Kingdom. We interviewed money laundering reporting officers, financial crime staff and staff on the front-line in the branches and call centres of the organisations. Both of the organisations had advanced systems for the reporting of suspicious activity and the procedures employees followed were very well integrated into customer management processes. There was a degree of homogeneity between the organisations as they had recently merged and were seeking to harmonise their AML/CTF processes. In spite of this climate of harmonisation, the remediation work accomplished by staff in these organisations was complex. Because of the decentralised nature of AML/CTF staff had to perform multiple remediations to comply with the regulations. Front-line branch and call centre staff remediated identity documents and potentially suspicious activities that they observed. Financial crime staff remediated intelligence from the frontline and government watch list information into the firms' securitised information flow.

Transforming media

The two tenets of AML/CTF, 'know your customer' (KYC) and 'customer due diligence' (CDD) directly implicate customer-facing staff. Each process involves staff remediating customer information for security purposes. KYC dictates that customers must undergo identity checks before being offered a product; CDD dictates that customers' accounts and behaviour should be monitored and screened constantly for anything unusual. It becomes the responsibility of front-line staff to satisfy both of these requirements. As we shall shortly demonstrate, and in contrast to staff in the travel sector, both of these regulatory processes were well integrated into the everyday work of employees. Research participants were swift to indicate that they had become responsibilised for AML/CTF. One participant noted '...at the end of the day, we're the first line of defence' (F25, Call Centre Employee, Company A) and another: 'I think that it's just unfortunate the society that we're in now. And if we don't do something about it no-one else is' (F28, Branch Manager, Company B). The following paragraphs describe each process in turn.

Both organisations had very similar KYC processes. After receiving an application for a product from a potential customer, the customer's information was entered into the CRM system which then performed a credit check on the customer. Following the credit check, the system dictated the identification documentation to be requested from the customer, which usually comprised a proof of identity and proof of address. Different types of document were required for each: a passport or driving licence provided proof of identity and a utility bill, rent book, bank, building society or credit card statement, among other things, provided proof of address.

These documents were verified by staff in branches or head office and then sent for processing so an account could be opened. If the documents were inadequate, for example, if names and addresses did not match, birth dates were inconsistent or a middle name was missing, further identification documentation was requested from the customer. This branch employee explains the process:

> When we see customers face to face we go through an application and they will get electronically scored, but they all get electronically scored and depending on what that scoring is people will have to provide more proofs of ID and address because the system has linked them to that address … And then what our guys do is photocopy them, stamp them, certify them to say that they're actually real valid copies, that there's a likeness to the customer sat in front of them, you've not got Kermit in front of you.
>
> (F22, Senior Branch Advisor, Company A)

While this employee begins by describing the process in bureaucratic terms, they then indicate how this system immediately begins to discriminate because of the norms surrounding acceptable identification and identity which are embedded in the credit scoring system used by the bank. The remediation work undertaken by front-line staff involves aligning the system's dictat as to what type of identification is acceptable in terms of the regulations, with the variety of paper-based identity documentation and the variety of human beings who present themselves as potential customers. Ultimately, this work, which involves vigilance and inspection on the part of employees, determines whether security-related information flows about the customer at all. Employee comments immediately venture into the use of uncomfortable stereotypes, although they were not used unreflexively, indicating the generative impact of the normative judgements involved. Comments about age and nationality emerged. The first two comments concern age:

> There are certain things in the AML that are pretty difficult – young customers. They might have a passport they normally don't have any bills because obviously the parents are paying the bills.
>
> (F22, Senior Branch Advisor, Company A)

And older people in nursing homes:

> ... want to open accounts or fixed term deposits or things like that, but they can't really provide proof because of the way they're living. They might have sold the house, or the family might have sold the house and they're in this nursing home, but obviously everything goes to the nursing home. So what can that person provide to be able to open an account?
>
> (F22, Senior Branch Advisor, Company A)

These two categories of customer warrant special attention in the JMLSG (2006) regulations as well. The following comments indicate a clear tension surrounding common national stereotypes currently experienced in Britain which cause the employee to justify greater vigilance. These are not referred to by the regulations, but it is clear that the presence of certain kinds of difference attracts a more intense gaze:

> I've had some really dodgy Nigerian ones, a couple of weird Italian ones, it gets to the point where you start, as soon as you get a passport from a certain country, you start thinking, hmm, you get some really dodgy ones at times.
>
> (F24, Financial Crime Employee, Company A)

This extra vigilance, which mobilises stereotyping which might well be legally problematic in terms of anti-discrimination legislation, is justified by commercial and customer services imperatives which are underpinned by regulatory compliance. Incidentally, in early 2014 BBC Radio 4's 'Money Box' detailed a court case brought by nine innocent Iranian citizens living in the UK whose bank accounts had been frozen using the money laundering regulations.[11] This branch manager explains:

> ... we are salespeople ... however it doesn't matter how good your sales are, your sales could be 200 %, if you're not managing your compliance ... again you would not get 'achieving' [reference to internal performance rat-

11 http://news.bbc.co.uk/1/shared/spl/hi/programmes/money_box/transcripts/money_box_04_jan_14.pdf accessed 9th January 2014.

> ing system], because you're not doing your job, regardless as to how good the sales are.
>
> (F28, Branch Manager, Company B)

Finally, another employee confirms that if a customer perhaps had nothing to hide, then they would have nothing to fear:

> And if they're a genuine customer, nine times out of ten, they'll thank you for vigilance. Because they think to themselves, if I was a fraudster, there's no way I'd get into this account.
>
> (F25, Call Centre Employee, Company A)

CDD activities took place at both the front-line and in financial crime departments. In the words of one financial crime manager, bank staff work 'to either flag them up at front-end or to do retrospective sweeps' (F27, Financial Crime Employee, Company B). The outputs of these processes resulted in new intelligence which was then remediated by financial crime staff into fraud systems for investigation. Whilst we discuss more about frontline staff and CDD in the next section, financial crime employees undertook CDD in response to data about their own customers and from government. As well as monitoring the transactions of their customers they were also required to run searches of their customer databases in response to new watch lists released from NCA, HM Treasury, Department of Work and Pensions, local authorities or the police. Such requests would concern funding evidence of persons suspected of being involved in organised crime or terror, or who were politically exposed. Financial crime staff complained that the volume of work that they had to do was dictated by the frequency and length of watch lists, both of which had increased in recent years. They called this phenomenon 'an evidential spike'. The financial crime managers in both banks were careful to set the parameters of the searches which resulted in a manageable amount of positive identifications which would then require further investigation. Smaller scale cases were dealt with by an outsourcing partner in India, whereas the more complex cases were dealt with in house. This MLRO explains:

> ... the trick is making sure you don't produce so many false positives that you're just blinding your analyst ... Because the risk is if you're sitting

> there just trailing through thousands of payments ... you're going to miss something because you're going to be blinded by the one that is a match.
>
> (F18, MLRO, Company A)

This MLRO goes on to explain the amount of internal monitoring that is also undertaken:

> We screen all of our customers on a daily basis. But we screen it at the day they open the account up, we screen it to see if it's anybody on the list. If they change any of their details, if they change address, or they change, say, name or anything like that, we'll re-screen them because that additional information may now match something we've not found before. Any time the details on the list change or are updated or are added to, we'll screen everything again. So we're constantly screening.
>
> (F18, MLRO, Company A)

In the final quote a financial crime employee explains how their transaction monitoring system identifies suspicious activity: remediation work takes place as the financial crime staff seek to establish genuine suspicious activity and then undertake further reporting. They highlight how different types of account attract different rules as well as highlighting the significance of cash withdrawals. On some types of account, high proportions of cash transactions can provoke suspicions about links to terrorist organisations.

> I think we've got about nine rules with regard to the savings accounts and five on mortgage accounts, and we'll go and look ... For example, one of the rules could be a large cash deposit or large cash withdrawal. If it happens on a customer's account, we would get an alert. That comes in every morning.
>
> (F26, Financial Crime Employee, Company B)

Reconfiguring resources

We now move on to discuss the second element of remediation work: reconfiguring resources. Under this heading we discuss how front-line and financial crime staff used the resources at their disposal: the customer, information systems, organisational rules and their own tacit knowledge to identify suspicious activity and then act upon their suspicions. As well as feeding the securitised information flow and this also enabled the employee to discharge their regulatory responsibilities. Two activities are critical: reassembling, where employees used embodied customer service techniques to bridge the gap between commercial priorities and security interests, and pre-empting, where employees used their tacit knowledge to second-guess the customer and identify any malign intentions.

Reassembling

Within AML/CTF practice the details of the customer service encounter are nearly always reassembled as a security-based encounter simultaneously or shortly afterwards if an employee has suspicions. We use the word 'reassembled' because of its specific meaning in the context of remediation work. Employees 'put together' a case for notifying the financial crime department of their suspicions as it unfolds in front of them. They rework and requalify what happened in the customer interaction and, finally, remediate their suspicions into the fraud system so they can then be acted upon. The notion of reassembling also invokes the alternative meaning of remediation, that of 'repair' – how elements of the working infrastructure are put together in new ways at particular times for regulatory ends. In many ways this is inevitable because of the fundamental gap between security and sales practices. It is particularly fascinating, however, that in spite of this gap, both the KYC and CDD elements of the regulations arise within the customer relationship. During that interaction, the front-line customer service worker is required to bridge the gap between service and security, monitoring the transaction for suspicious behaviour. But the employee must do this without the customer realising until, in the words of a financial crime employee: 'the police knock on the door if they feel it's necessary' (F26, Financial Crime Employee, Company B). If they do, the employee can be accused of 'tipping off' which is an offence under the law that underpins the AML/CTF regulations, the Proceeds of Crime Act 2002. If it was subsequently discovered that a staff member had not reported suspicions of serious crime, or had tipped off a fraudster, they could receive at best a fine or at worst a five year prison sentence. One participant said this prospect was 'frightening' for staff (F28, Branch Manager, Company B). As such, in both organisations, front-line staff were told to be vigilant and suspicious at all times and not to take customers at face value. In spite of this, financial crime staff noted that front-line staff were perhaps over cautious in their reporting which added to their workload. A financial crime manager explains tipping off in simple terms:

> ... if somebody wanted to pay £5,000 in and we said to them, "oh has this come from straight from the bank?" And they turned round and said "it's none of your business, I'm not telling you where it's come from". They [the employee] couldn't then turn round and say "well I'm not paying that in, because you're not telling me where the money's coming from". They would never be able to do that because that would arouse suspicion, it could be classed as tipping off to the customer. So they would just carry on with the transaction, they would do that, they would probably make a note, a discreet one, once the customer had gone.
>
> (F28, Branch Manager, Company B)

The issue of 'tipping off' highlights the irreconcilable gap between security on the one hand and sales and service on the other. In the incident described above a performative veneer of customer service conceals the front-line staff's reassembly of the service encounter into AML/CTF evidence 'on the hoof' so it can be fed into the AML/CTF information flow. A bank call centre employee describes the same process as it plays out on the phone. She uses the information she sees on her screen, combined with language concerning bureaucratic processes and her background knowledge of organisational procedures to keep the customer comfortably in suspense:

> You don't let on in any way that you're suspicious. You just process as normal. Say, okay, that's fine, here's your application number. One of three things, it'll either decline, it will either go through accepted, which means it's agreed, so just around formalities, or it will refer, needing the further information ... Let's say they come back in a couple of days to get a decision, it could be that there are notes from fraud on there, from AML, because they're checking things. And we'll say, you just have to wait a couple of days, it's still coming through, we're still waiting for the result from the team. Not letting on in any way that we're checking into them, basically.
>
> (F25, Call Centre Employee, Company B)

Similarly while front-line staff are not supposed to reveal their suspicions to customers, financial crime departments sometimes do not feed back to front-line staff on their reports in case it influences how front-line staff then deal with the customer. For financial crime departments it is vital that the customer remains unaware and the service veneer is preserved. Chains of secrecy begin to characterise interactions between customer, frontline staff and financial crime departments. This branch manager explains the response of the financial crime unit when she asked what happened as a result of the reports she had submitted about a customer:

> they said you won't get feedback; there's two reasons you won't get feedback. One, because if you do get feedback you might treat that customer differently, and that's classed as tipping off, and if the customer gets in tune with that obviously then you're again tipping him off when you might have not meant to.
>
> (F22, Senior Branch Advisor, Company A)

A financial crime analyst also explains the response:

> we understand a lot more than the front-line people will, we still have to make our own decision as to whether it is suspicious; whether we feel that that could do with looking at in a bit more detail in a couple of months' time after we've got a bit more of an idea, their account use, or it's just not suspicious and we can close it off and just wait. It might crop up again, and then we can take the bigger picture.
>
> (F23, Financial Crime Employee, Company A)

Branch and call centre employees also illustrate the secrecy point. The first extract below is a comment made after a branch employee had described the levels of performativity required when a known fraudster had entered their premises. Whilst keeping the customer at ease – offering them tea, biscuits and a comfortable seat while they 'processed the transaction' – branch employees quickly assembled information from their fraud, prosecutions and security departments, so that they could call the police and have the person arrested on the spot. An employee who was involved noted, in a very matter of fact way, that 'dealing with people 24/7 means actually you're quite good and quite adept at having those kinds of conversations' (F22, Senior Branch Advisor, Company A).

Further, this call centre employee describes how they use the art of conversation to draw information from the customer on the phone:

> ... naturally, after being here ten years you pick up something in your first five seconds: this doesn't seem right. So then, without alerting them we just go, okay, can I just confirm more things for you? By doing that, you're listening to what they're saying and you're analysing what they're saying ... you're not trying to trip them up, but you're trying to see if they are who they say they are, rather than trying to move on.
>
> (F25, Call Centre Employee, Company A)

Both speakers begin to highlight the second element of 'reconfiguring resources' which forms part of remediation work. That element concerns how employees build up tacit knowledge which enables them to pre-empt when something is amiss in either a customer interaction or transaction. We now discuss this in more detail.

Pre-empting

The importance of front-line staff in identifying suspicious activity has already been highlighted in previous chapters. In particular, we referred to a comment by a money laundering reporting officer (MLRO) of one of the financial services organisations in the study who revealed that only 40% of the suspicious activity they reported was identified by data analysis. The rest came from the front-line. Front-line staff confirmed this view. It appeared that judgements about the customer's demeanour, their comportment and attire, the circumstances of their arrival and departure from a branch, or whether 'things added up' about them or not, mobilised the stereotypes of suspicion cemented in the tacit knowledge of front-line staff. One factor was:

> if the customer is ... getting a little bit nervous or they're getting a little bit standoffish they may be thinking there's something not quite right.
>
> (F22, Senior Branch Advisor, Company A)

Details about their income and occupation may not make sense:

> it could come through saying they're a waitress and they say they earn £50,000, for example. It's happened. So we've then got to do further investigations.
>
> (F25, Call Centre Employee, Company A)

Contextual factors in the branch also contribute to suspicion, as well as whether one has a visible body modifications, which presumably are being read as an indicator of deviance:

> the customer has come in, he's drawn £4,000 and what he's said it's for is a new car. However, outside … he went back inside and got in a Mercedes that was 2011, brand new, it was like this was the registration plate, the customer had a tattoo on his left arm …
>
> (F22, Senior Branch Advisor, Company A)

Language is also an issue. This branch manager describes a situation where a newly arrived immigrant who can't speak English is trying to open an account. Eventually the obvious language barriers cause the manager to become suspicious:

> he can't speak English so how can I physically open you an account. That's difficult because you might have his brother sat next to him who has brought in four people in one week – this has happened before – and he's trying to open an account for these people because he is doing the translation for his family. How do you know that he's telling that person what you're telling him?
>
> (F22, Senior Branch Advisor, Company A)

Another relates how a simple hunch might cause the employee to question the validity of ID documents:

> I don't like this, there's something not right, the customer doesn't fit, maybe the passport looks … it might be fake because we've had that before.
>
> (F22, Senior Branch Advisor, Company A)

This was also true for those looking at financial data. Most significantly, one of the clearest benchmarks of whether a customers' activity has aroused suspicions or not is the organisation's own product portfolio. Through the use of CRM, the organisations in the study had developed a range of products to suit different lifestyles. Whether a customer appeared to be conforming to that lifestyle and the range of financial transactions that accompanies it was enough to raise questions:

> ... it's all about lifestyles, and obviously different accounts are for different lifestyles. Because obviously a very basic account would be for someone who may not be earning that much, hasn't got enough of a credit report to get a higher level account. So really, although you can't use it as law, really, though, looking at one account can let us know or give us an idea of what we shall be seeing going through that account, as compared to someone else.
>
> (F23, Financial Crime Employee, Company A)

Conclusion

Remediation work is a central element of the AML/CTF which enables both the front-line staff and the institution's financial crime department to feed the AML/CTF securitised information flow with new information. A number of entities were transformed into digitally mobile information as part of the AML/CTF reporting process. Under KYC, customer identity documents were verified and checked against system requirements, so the customers' identity becomes remediated as 'acceptable' or otherwise. The remediation work associated with it involved aligning the identity document, the potential customer and the system which assessed their risk. Under CDD suspicious customer transactions and customer behaviours become re-mediated into fraud systems. This aspect of remediation work took place both at the front-line and in financial crime departments. Financial crime staff found this a source of work intensification as the volume and pace of their work was dictated by government requests, their system settings and the vigilance of front-line staff. At the front-line, the transformation of media for regulatory ends seemed to be underpinned by a politics of bureaucratic normativity based upon the risk associated with different identity categories. Also apparent is the presence of market logics which justify the scrutiny of customers for commercial *and* regulatory ends as a win-win. The differences between commercial and regulatory interests, however, were highlighted when the resource reconfigurations around KYC and CDD were examined. Employees worked hard to formulate their reasoning about their suspicion to render it 'reportable' in AML/CTF systems (reassembling) and used their experience of dealing with customers to tease that out (pre-empting). But these activities have deeper significance too. They signified that the variety of human beings which present themselves to front-line staff are subject to a range of tacit judgements – some of them grounded in financial data, some grounded in marketing data and product placement, others grounded in 'hunches' – which mobilise suspicion. Whether things 'don't add up', 'or don't seem right' about a customer seems to be embedded in organisational folklore, learned through experience and passed on between colleagues through training, coaching and feedback processes. It is this range of judgements which initiate securitised information flows in the financial services sector. And the threat of sanction, security and reputational breach will ensure that these suspicions are always in circulation. We now proceed to discuss remediation work in the travel sector before drawing some comparisons.

Comparing the two cases

The statistical analysis and qualitative data presented in chapter 6 suggested that AML/CTF had a huge impact on customer-facing staff and posited that the re-focus on the customer relationship in retail travel would have the same effect there. By using the newly-developed lens of remediation work and its sub categories transforming media and reconfiguring resources, we have explored this impact in depth and revealed the type of work that occurred. There are some differences between the cases which we will discuss first, but significantly, there are some striking similarities too. The first and obvious difference is that more complex local re-mediations took place in financial services organisations whereas in travel, simple, local re-mediations of passports were then transferred along the supply chain to airlines and eventually to government. Second, because the identification of suspicion was decentralised in financial services this resulted in some interesting 'rules of thumb' and stereotypes about what constituted suspicious behaviour. This was not the case in travel, where information was transferred en masse. In terms of remediation – work, however, a critical difference was that 'integration' remediation work did not occur in the financial services case study. This was due to its maturity as a scheme and the fact that its practices are well embedded into job descriptions, procedures, training manuals, systems and performance management processes within the organisations we studied. As such the financial services front-line did not experience work intensification in the same way that the travel sector front-line did. Examining the experiences in the travel sector revealed that much of the work intensification arose from the efforts required to integrate eBorders into everyday working practices. Nevertheless financial crime employees did experience work intensification in a way which is comparable to that of travel sector front-line employees. The former's work was intensified as governmental demands for checking and screening against their watch lists increased. The latter's work was intensified by the mere presence of a government demand for passport information. In other words, direct governmental demands for information increased the workload of select groups of employees in the organisations concerned in a non-negotiable way. Other similarities also occurred:

Alignment of hybrid entities: In both cases remediation work involved the alignment of a wide range of hybrid entities to ensure that a working infrastructure was created and securitised information flowed. In financial services, a working compatibility between a host of elements was achieved: knowledge of customer behaviours and organisational requirements were combined with transaction data, customer relationship management and fraud information systems, government lists, data search practices and outsourced organisations as well as the legal elements of the scheme to create the flow. Similarly in Travel, the elements included the customer, their passport, proprietary information systems and those of different airlines and other businesses in the supply chain.

Reassembling and pre-empting as key activities: it was striking that the remediation activities of reassembling and pre-empting arose in both cases. Reassembling refers to how employees put different resources together in new ways to enable security information to flow upwards in the organisation and out to government agencies. Pre-empting refers to how employees used their tacit knowledge of customer behaviour to create (financial services) or extract (travel) that information. We argue that both of these activities arose as a result of the gap between regulatory duty and commercial priorities. Front-line employees worked hard to reconcile these competing interests and a lot of the time the nature of that work was unacknowledged by their employers. This analysis shows that in particular the embodied knowledge and ongoing judgements made by front-line workers positions them as pivotal in both regimes. This was particularly apparent in the financial services sector whose front-line employees relied on 'hunches' to probe potentially suspicious customers as well as convincing performativity when their suspicions were confirmed. In the travel sector agents and operators spent time second guessing their customers inability to enter data correctly or remember to submit passport data in the first place.

Market Logics as a compliance rationale: Interviewees in both sectors used market logics to justify their particular response to the regimes, although that response was often quite different in character. For financial services employees, compliance was positioned as a 'win-win' through achieving 'the compliant sale'. In chapter 6 we outlined the reputational and strategic benefits of compliance which underlines this particular stance. For staff in travel firms, who were still adapting to the regulations, 'turning eBorders into a sales opportunity' and 'not losing the customer to the competitor' were strong compliance rationales expressed by those in strategic or managerial positions in the firm. However for those front-line staff trying to fit eBorders around their other job responsibilities the advantages were less clear. The mobilisation of these logics in practice is at the heart of remediation work: embodied knowledge, judgements, technologies and other resources recombined to ensure that data flowed while market priorities were maintained.

Responsibilisation, anxiety and fear: In both cases employees acknowledged that they felt responsibilised by their respective government surveillance regime: in other words, they felt that they had a duty to discharge their regulatory obligations and expressed a sense of anxiety, stress and sometimes fear if they didn't. In financial services, some of the employees we interviewed actively – almost heroically – embraced that responsibility. However the threat of sanction for making a mistake was very real for all involved. One employee even described the prospect as 'frightening' and that it had caused the staff to become over cautious in their reporting of suspicion. This is a phenomenon that has been observed industry wide and is reflected in the huge jump in the number of SARs submitted to NCA once the regulations were changed in 2007 to criminalise bank employees for non compliance. Although firms in the travel sector were threatened with criminalisation for non-compliance with eBorders this threat had not been passed on to staff. As we discussed in Chapter 2 nobody has been fined under the regime because of its questionable legality in Europe. As such employees in the Travel sector were less responsibilised by compliance *per se* but they felt more responsible for ensuring that the customer was able to travel because of the threat to reputation and repeat custom if they did not. As we discussed in Chapter 4, for the travel agent, eBorders became a threat to commercial survival, especially for travel agents, because it had reworked priorities in the customer relationship.

CHAPTER NINE

The private security state

Responsibilisation, surveillance and security

Introduction

In this volume we have explored what happens when private sector organisations become enmeshed in government-driven national security regimes. We have examined this in the light of practical developments concerning the delegation of regulatory responsibility, the use of private sector partners to deliver government services and the outsourcing of security provision to the private sector. We have also considered this against an interdisciplinary theoretical backdrop of surveillance and security diffusion, wider normative considerations concerning the role that businesses *should* play in society and political revelations depicting how communications data from mainstream service providers has been accessed and used by security services.

We examined the experiences of different stakeholder groups, specific organisations and their employees in two industrial sectors – retail financial services and retail travel – as they responded to and complied with two British surveillance regimes which require companies to gather and analyse consumer data for security purposes. We found that industry stakeholders were unevenly affected by the regulations because of their differing interests and power in the stakeholder network, but that they each experienced them as a compromise to their commercial operations. We also found that as firms adapted to these regulations, they did so in such a way that the regulations were subsumed under market logics and used as a means by which business advantage could be leveraged. In both sectors, this resulted in reworking regulatory requirements into the customer interface in a way that it could be justified as adding value. However this process of reworking resulted in those at the customer interface: call centre workers and customer-facing workers in branches, absorbing this extra security work as a form of invisible work for which no additional recompense was received. This new form of work is termed 'remediation work'. In this chapter we reflect on the end point of this research, its contribution and what needs to happen next. We begin by considering how we have developed thinking on the political economy of surveillance, regulatory compliance, thinking on the broader responsibilities of business and for the sectors themselves. We then offer some reflection on our own research process for those engaging in this kind of work in future.

Surveillance theory: The mid-range and the securitised information flow

At the beginning of this book we noted that the range of published empirical work so far on the phenomenon of surveillance has been polarized. Wide ranging commentaries (Lyon, 1994, 2001; Gandy, 1993, 2010) sit alongside detailed empirical pieces examining local practices (Norris and Armstrong, 1999; Smith, 2007; Neyland, 2007; McCahill and Finn, 2014). The current body of work on surveillance is, however, missing a consideration of how the structures and practices implicit in this body of work may be connected. Various theoretical works (Ball, 2002; Haggerty and Ericsson, 2000) which address this issue focus on system components rather than what brings them together.

We approached the analysis of eBorders and AML/CTF by focusing on how organisations went about creating a securitised information flow as a result of surveillance-based regulatory interventions. Our contribution to surveillance theory is to move beyond macro level commentaries and micro level analyses and augment this midrange as containing numerous entities which had to come together for surveillant ends. These are entities which go beyond the obvious 'technical' focus which is often the first port of call in analyses of surveillance systems. Although information systems are important, we also highlight how the roles of industry standards, financial resources, the embodied knowledge of front-line workers, organisational routines and working practices, intangible assets such as brand and reputation became enmeshed in the production of surveillance. We have argued elsewhere that surveillance is an organisational phenomenon (Lyon, Haggerty and Ball, 2012) and here we demonstrate exactly the range of organisational elements involved. This work augments Haggerty and Ericson's (2000) discussion of the surveillant assemblage in this respect and helps to populate the 'elements of surveillance' identified by Ball (2002). Ball (2002) argues that surveillance practices have a number of interconnected elements: (Re)presentation, the material technologies and infrastructures which carry and present surveillance data; Meaning, the social, cultural and organisational dynamics which shape the manner in which those data are interpreted; Manipulation, the presence of power relationships and the pervasion of interests which determine whose meaning counts and

Intermediation, the arrays of actors which bring the different elements of surveillance together to create information flow. This work highlights the complexities of surveillant Representation and the vagaries and tensions inherent in infrastructure development. Meaning making in the cases relates not only to local workplace cultures and practices within the organisation but the reputations of global corporations and interpretations of customer behaviours outside it. Manipulation stems from the fundamental divergence in public and private interests as well as the uneven distribution of power between industry stakeholders. Intermediation rests on the assembly of organisational elements brought together under market logics which are defiant to government attempts to colonise the private sector with public interest matters. Instead the private sector seeks to re-colonise the public interest realm of security with commercial opportunity. Multiple layers of adaptation took place in order for information flow to be created and organisational elements became aligned in new ways: ways which were driven by market logics rather than compliance-based ones. Most critically, remediation work is at the very centre of what brings these elements together. It is the stuff of the surveillant labour process and initiates securitised information flows. This is what constitutes the mid-range of the two surveillance regimes we investigated. This is what allowed information to flow from the surveilled subject - the customer - to government.

As this information flow was constructed and sustained, the surveillant gaze it carried became deflected and refracted through the mid-range and its constituent parts. There are some clear reflections here on the nature and direction of the surveillant gaze as we observed it which critique the ideal-typical way in which Bentham's Panopticon (Foucault, 1977) is often referenced in Surveillance Studies. Rather than a singular gaze, we observe a multidirectional, opaque and fractured gaze carried by an information flow which is locally disruptive. The tensions and competing pressures which emerged seemed to reflect Latour's (2005) observation that each public and private domain which features observation and surveillance constitutes a series of oligoptica, the connection of which is fraught with mismatches, holes and work-arounds. Returning to Foucault (1977), however, we also observed how the construction of an information flow enacted power relations between the respective government agencies and the private sector supply chains from which they gathered data. Rather than resulting in the dominance of one interest group over another, competing pressures and tensions produced private sector re-appropriations of government surveillance processes and a re-subordination of them under market logics. Simultaneously, governmental authority in national security matters was enacted through the threat of fines and sanctions for non-compliance. Local power relations were reinforced, with the exploitative social-economic relations of the private sector firm intact, as well as governmental authority.

In our cases, 'governmental authority' for national security related to the government agencies of NCA and the UKBA respectively. Initially we viewed these entities as 'stakeholders' and examined their accounts through a stakeholder theory and information infrastructures lens in chapter three. We outlined how the instantiation of information flows precipitated cost, technical and contextual tensions within different stakeholder groups. The prevalence of these tensions, for us, highlighted the challenges inherent in the alignment critical to creating securitised information flows. The tensions observed were described in terms of a challenge to the legitimacy and autonomy of organisations' ability to conduct business in the way that they wanted to, indicating a gap between security interests and commercial priorities. They also promoted new tensions and interconnections between existing stakeholders, prompting them to collaborate in new ways, creating new organisational and pan-organisational orders in order that these gaps could be overcome. The remediation work undertaken by front-line customer facing staff in both sectors, and financial crime staff in the financial services sector, involved not only integrating regulatory priorities into the customer interaction but reassembling resources around that interaction to ensure that data flowed.

In the travel sector, airlines were required to transfer passport data *en masse* relating to every passenger they carried to the UK Border Agency. Complying with these requirements had impacts all the way down the travel supply chain as supply chain partners became linked in to the eBorders information infrastructure. As there were multiple points of entry into this infrastructure – mirroring the multiple points of sale for airline seats – organisations dealing with multiple airlines, such as travel agents and tour operators, were hugely impacted. Each firm went through a process of recognising, rationalising and refashioning eBorders requirements into their business models. They worked eBorders into their customer service approaches and used it to improve the service experience and to cross sell other travel products to the customer. However the impact on staff was such that they incorporated extra remediation work into their tasks, as they transformed passports into data and reconfigured resources to ensure that this happened. Their use of tacit knowledge surrounding customer behaviour and existing information infrastructures was extensive. We summarise the travel case in Figure 9.1.

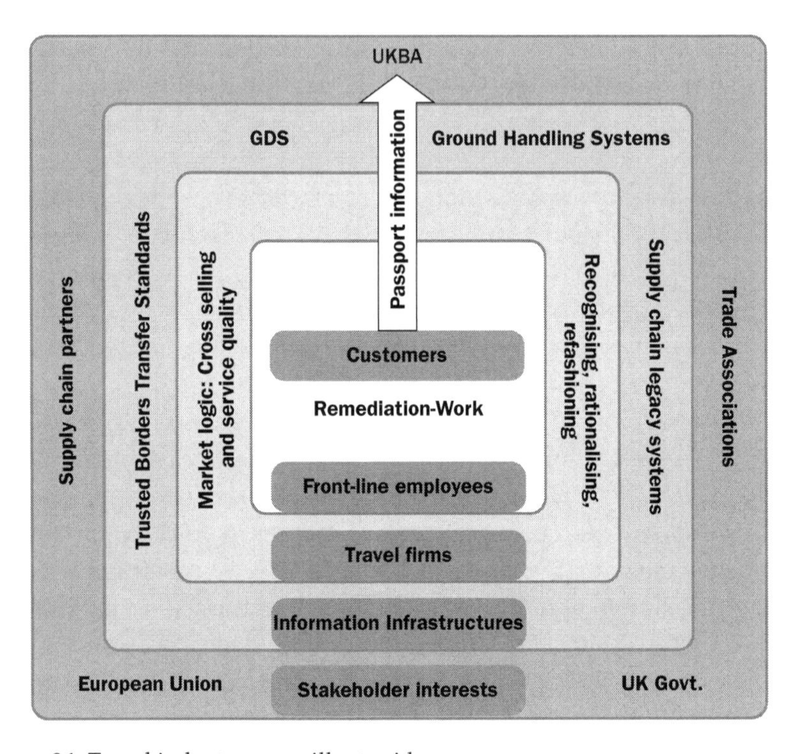

Figure 9.1. Travel industry surveillant mid-range.

In the financial services case, financial institutions were required to risk-analyse data about their customers, their transactions and use of financial services products. They then transferred the results of that analysis to the National Crime Agency in the form of Suspicious Activity Reports. Because the institutions analysed data locally, the impacts were felt mainly within the institutions and did not affect supply chain partners. At the time the research took place, AML/CTF information infrastructures were well established and comprised a single web portal on the NCA website as well as the institutions' internal systems. The regulations were fully understood in our case organisations and they were a routine part of banking work. However, at a strategic level, questions remained over the costs associated with compliance although the potential impacts of non-compliance on the organisation's reputation were acknowledged. As such the market logics which surround achieving 'the compliant sale' emerged, which addressed both commercial and regulatory ends. Parallels between CRM and AML practice were identified in our survey data, but in practice there was little official knowledge sharing between CRM experts who focused on the customers who were attractive to the organisation and AML experts who focused on the customers who weren't. Although internal AML systems were able to identify some suspicious transactions through algorithmic surveillance, front-line staff had a major role to play as they interacted with customers on a daily basis. Through the performance of remediation work, particularly by mobilising their tacit knowledge of customer behaviour they were tasked with identifying those who were suspicious as well as selling financial services products. This is summarised in Figure 9.2.

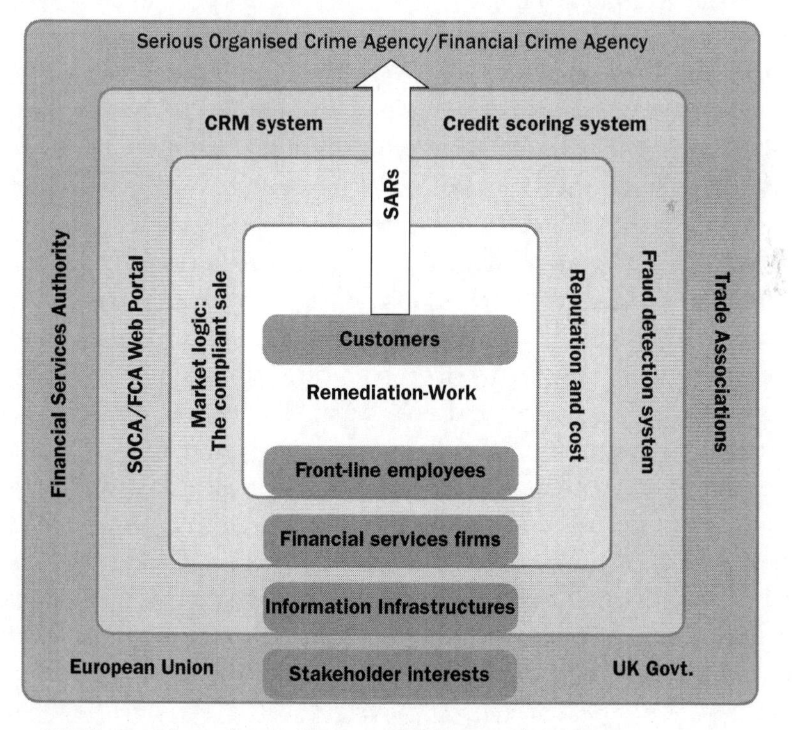

Figure 9.2. Financial services industry surveillant mid-range.

Remediation work: a new concept for an age of diffuse security?

As Loader and Walker (2010) have argued, governmental security activities are now diffused through numerous social and organisational settings to form a hybrid security landscape. Huysmans (2011) has pointed out that 'little security nothings' pop up in the everyday lives of workers as the state extends its reach into and annexes the private sector, using it as a data-supplier. Although customer-facing employees in both sectors are not primarily security specialists, they are utilised in a data gathering capacity by a national security regime via their employers which are deputised to it by law. 'Re-mediation work' is an important new category of security work as the employee transforms the customer into the subject of a security regime by re-mediating their identification data, information about their identity, behaviours and transactions (financial services) or their passport data (travel). One of the concept's strengths is that it enables concrete work activities to be understood in a context which spans beyond the employer to other institutions via pan-organisational information infrastructures.

Although remediation work was fully integrated in the financial services case, travel employees characterised it as mundane, which resonates with published understandings of security work (Wakefield, 2003). The tensions invoked in travel through 'integrating' re-mediation work, where employees tried to combine eBorders data capture with their everyday work, also resonate with the dialectical relationship predicted between commercial priorities and security priorities. Also in travel, re-mediation work was viewed as obstructing job priorities, resulting in anxiety and frustration. Through 'reassembling', workers in both cases created a bridge between the customer as a customer and as a potential security risk, further highlighting this dialectic. Particularly in travel, but also in financial services at times, in reassembling resources employees repaired misalignments in the infrastructure, enacting the important double-meaning of re-mediation work, 'repair' work, by enabling passport data to flow (Graham and Thrift, 2007). This entailed bridging infrastructural gaps, inventing new processes and documentation and reassuring customers while they did so. Through 'pre-empting', the resolution of this tension is partly achieved as employees applied their tacit knowledge to help them deal with regulatory requirements.

This is also where a new politics of surveillance-based security regimes arises. In Chapters 7 and 8 we exposed how re-mediation work represented work intensification, resonating with studies of public–private partnership which identify declining labour conditions as a problematic outcome of the process (Grimshaw et al., 2002; Taylor and Cooper, 2008; Howcroft and Richardson, 2012). Of particular salience are Flecker and Meil's (2010) observations that those with the least bargaining power gain the least from such activities: there was little or no room for negotiation of re-mediation work in any of our research sites. In financial services, remediation work had become normal for front-line workers, but financial crime specialists were subject to changes in governmental demands and had no choice but to respond. In the Travel sector, customer-facing employees were the least powerful stakeholders on the edge of eBorders infrastructures. Yet they played a pivotal role in enrolling the customer into a security – relationship with the state as they purchased and consumed a travel product. Huysmans (2011) argues that security becomes de-politicised as it is diffused. However we argue that its diffusion through to the private sector results in a resurfacing of this politics in the politics of production. We have shown that the intersection and integration of security issues into the everyday work roles of some of the lowest paid clerical workers in the service economy requires extra – often unrecognised – effort. And as market logics dominate compliance rationales, this extra effort is deployed not only to the end of security but also to the ends of maintaining commercial reputations, customer satisfaction, profit margins and hitting sales targets. Although de Goede (2012) points out that pre-emptive surveillance such as this is technical, non-violent and speculative, it remains exploitative because of its location in the intensification of front-line service work. Re-mediation work can be applied to numerous situations in which information is re-purposed and re-packaged for security ends. Recent examples include the current UK government's Prevent Strategy (Home Office, 2011), which calls for heightened vigilance in combatting terrorism by universities or city councils and the Data Retention and Investigatory Powers Act (2014) which involves communications service providers.

Surveillance, regulatory compliance and regulatory capitalism: Changing the business of business

Our findings also create insight into processes of regulatory compliance from the organisational perspective, which are under-reported in the literature (Parker and Nielsen, 2009). We identified both AML/CTF and eBorders as symptomatic of the growth in 'non-state regulation': the use of non governmental actors to achieve regulatory ends (Hutter, 2006). It was acknowledged in this literature that such regulation would have an effect within firms, but that this effect was likely to be negative in terms of competition, innovation and creativity (Porter, 2011). Writers such as Coglianese and Nash (2006), for example, speculate as to the form regulation may take and outline the idiosyncratic and culturally grounded responses by firms. The wide-ranging commentaries of Brathwaite (2008) and Levi-Faur (2005) identify the commercial advantages and disadvantages which stem from such regulatory changes as the state 'fills out' rather than 'hollows out' under the state-corporation win-win arrangements of regulatory capitalism. They argue that larger organisations with the capacity and resources to respond to regulation will experience competitive advantages as smaller organisations who cannot absorb the costs are less able to compete (see also Hanlon, 2008; Hanlon and Fleming 2009).

Our treatment of the securitised information flow contributes to this debate. In Chapters 5 and 6 we argued that over different timescales, firms in both sectors engaged in processes of recognising their regulatory burden, rationalising what was required and what would sit in commercial priorities and then refashioning aspects of their operation, particularly around customer contacts, to be compliant. Critically, securitised information flows to be used for the purposes of government surveillance are generated out of existing organisational processes which are already in train and are then applied to a completely different set of priorities. And it is these processes and priorities that support the dominance of market and commercial logics which dictated how national security surveillance was achieved in an everyday sense. One sector, retail travel, ultimately decided to embed data capture for eBorders within customer contact processes, having initially decided to rely on customer self-service which caused more problems than it solved. Once infrastructures were in place to transfer the data, it was the least powerful stakeholders in the supply chain – travel agents and their customer-facing staff – who had to do the most legwork to gather passport data from customers because they were afraid of losing those customers to competitors. This situation emerged because of the shape of the regulation as it affected the retail travel sector. With no single standard site of data input firms in the supply chain had to resort to a decentralised solution with companies making their own arrangements for data collection and transfer. The additional burdens caused by the regulation drove a wedge between legacy airlines listed on Global Distribution Systems and the retail travel supply chain – comprising tour operators, charter airlines, high street and online agents, 'screen scraper' providers and seat brokers. Retail financial services had clear guidance from the regulator as to how the AML/CTF regulations should be addressed. Unlike firms in the travel sector, who had to gather data *en masse*, financial services firms were to analyse customer data for suspicious activity and were to report it to NCA via a centralised portal. By allowing the financial institutions to determine how they were to respond to the regulations in terms of identifying the areas which were of greatest risk, AML/CTF gave rise to internal tensions in the firm between regulatory and business priorities (also documented by Amicelle, 2011). Whilst travel firms initially tried to keep eBorders away from commercial processes, financial crime specialists and those carrying the regulatory burden within financial services organisations had the task of integrating AML/CTF into customer management processes. In spite of attempts to automate the identification of suspicious activity, customer-facing staff were deemed most able to identify suspicious behaviour and they used their tacit knowledge to reconcile gaps between security and service.

We also observed anecdotally the competitive effects of regulatory capitalism in both sectors. Key informants in both sectors were clear that smaller firms were less able to respond to both sets of regulations because of the cost burdens that they imparted. This is particularly clear in the retail financial services sector as the larger banks produce the vast majority of SARs, able to capitalise on economies of scale and invest in large scale systems and high numbers of staff to investigate accounts. Smaller organisations who are subject to the regulations, such as solicitors or estate agents, find reporting considerably burdensome by comparison. Firms also begin from different starting points with these regulations: in financial services, for example, we argued that a firms' CRM orientation might impact their ability to respond to AML/CTF. In travel it was quite clear that eBorders posed a threat to ownership of the customer relationships of tour operators and travel agents. We argued that there were potential long term structural outcomes of this regulation for retail travel as it was altering the terms of competition both within and between levels of the supply chain.

It is therefore clear that these regulations impacted multiple organisational levels, shaping the conditions of possibility for securitised information flows. We charted impacts on information infrastructures, stakeholder relationships, strategies, approaches to marketing and the experience of employment. These different elements of the organisations were responsibilised for national security matters in different ways by either building, constituting or maintaining the securitised information flow. Whilst it might be tempting to say that the experience of those in the organisations we investigate would be the same no matter what manner of regulation was applied – environmental, food safety or tax, for example – the fact that it is security regulation, we argue, gives it some unique features which give rise to a new politics of responsibilisation and a political economy of surveillance.

Enacting danger: The new politics of responsibilisation and the political economy of surveillance

In Chapter 1 we drew on Huysmans' (2014) latest work to suggest that making commercial practices objects of security would enact them as dangerous. When we reviewed the data it became apparent that those on the customer services front-line in both sectors experienced levels of fear and anxiety because of the new responsibilities they had under the AML/CTF or eBorders regulations. This was expressed as a direct result of experiencing the responsibility for collecting security data to feed the information flow. In retail travel this fear arose because of the threat of the customer being refused travel if the employee had made a mistake in inputting passport data. In financial services this fear arose from the criminal punishments threatened if they failed to spot or report suspicious activity. The reason that these fears arose, we argue, was due to the fact that they were primarily dealing with security matters. AML/CTF and eBorders create new, local vulnerabilities to which organisations and their employees need to respond, and respond without question (Huysmans and Buonfino, 2008). In simultaneously compromising commercial priorities and supporting national security strategies, AML/CTF and eBorders place organisations and, particularly, their front-line employees who are framed by these regimes, in double jeopardy. A failure to address regulatory requirements potentially enhances national *in*security and could result in commercial insecurity as well. Local circulations of fear, anxiety and responsibility arise because of the risks associated with getting it wrong. And as private sector organisations, the response is for all to work harder in the relentless pursuit of profit. Profit must conquer the regulatory burden. As such, our analytical category 'remediation work' serves as an important vehicle through which the hidden costs of national security regulation can be foregrounded. Insecurity was the initial driver, surveillance was the response, but local insecurity and heightened responsibility was the result. These senses were intensified by a lack of feedback from the regulator to the organisations as well as within the organisations themselves, as to the effectiveness of the reporting. In financial services in particular we observed chains of secrecy emerging because of the danger that the customer-facing employee might tip off the customer. Bureaucratic isolation within these regimes heightened the sense of responsibilisation and insecurity.

This analysis also supports suggestions that a political economy of surveillance, featuring new combinations of the public and private sectors to promote the analysis and capture of data on the general population, is emerging in these industries (Ball and Snider, 2013). Highlighting the dynamics of the securitised information flow which feeds these government surveillance regimes has both a political dimension and an economic dimension. Its political dimension refers to how it embodies different interests, privileging some over others. Most notably government is positioned as the most powerful stakeholder as well as the principal organisation through which information is transferred. In eBorders the Trusted Borders consortium were also similarly powerful in that they dictated the shape of the scheme to other stakeholders in a way which significantly disadvantaged retail travel. In both sectors, larger companies with central, rather than peripheral, roles in supply chains were much better equipped to respond. Its economic dimension refers to how those interests become enacted through economic means, being subsumed under market logics and have economic consequences because market structures are altered by their presence. Earlier in our analysis we highlighted, in a manner similar to that which has been reported in the public–private partnership literature, how less powerful stakeholders lost out as a result, namely labour through work intensification and those less powerful companies at the end of the supply chain, such as travel agents. We are also reminded of Hanlon's (2008) critique of corporate social responsibility (CSR). Although AML/CTF and eBorders are not CSR initiatives in that they are not voluntary there are some parallels to be drawn. Hanlon (2008) argues that many firms see CSR as an opportunity to marketise new areas. A large multinational might enter into providing products or services which promote care for the elderly if it can generate new revenues providing them such products and services, or may manipulate bio-diversity or fair trade to its advantage. We have observed the same phenomenon in that firms struggled to regain authority and legitimacy in the face of these regulations by embedding them within the customer relationship: where the firm engaged with the market. The pursuit of the 'compliant sale' was hugely important in the financial services sector; the timely accurate recording of passport data, whilst selling extras and improving the travel experience was similarly so in travel. The political economy of surveillance with its diffused and depoliticised security practices therefore ensures that government driven surveillance practices are re-embedded in the politics of production. As security practices become subject to the politics of production they become drivers for the intensification of work and in particular forms of invisible, immaterial work (Hochschild, 1983; Lazaratto, 2006; Hardt and Negri, 2000; Star, 1999). Similary, even though Loader and Walker (2010) argue convincingly that security is a public good, through its subsumption under commercial aims, a reverse colonisation took place as this public good slowly became colonised by the aims of private capital.

Business in society: Cycles of regulatory response

Referring to Hanlon's (2008; Hanlon and Fleming 2009) critical work on Corporate Social Responsibility also reminds us of some of the debates reviewed earlier in the book which stem from business perspectives on the roles it should take in society. In particular, we referred to some recent work by Porter (2011). In considering the effect of regulatory transfer on the operation of private sector firms, he cautioned against over-determination by regulators. He argued that this could 'block[ing] innovation and almost always inflict[ing] cost on companies' (p. 74). Nevertheless it is clear that across the security landscape the private sector is becoming more closely involved with national security provision. Discussions by Loader and Walker (2010) and a host of empirical evidence document the spread of the 'new political economy of security', characterised by a broadly spread hybrid network of security providers where the state is now no longer seen as the sole source of national security provision. This 'public–private blurring' has also been observed within the development of the surveillance society (Surveillance Studies Network, 2006) especially around government service provision but also with respect to retail conglomerates who command huge datasets on consumer behaviour; and industries such as call centre, logistics and gambling who lead the way in employee surveillance.

In considering how business might respond to the demands of a wider social role, particularly if mandated to do so by a regulator, we note that levels of strategic flexibility and reflection on core business priorities are required. Given the long-term timescales within which regulation tends to operate and its tendency to transform rather than disappear (Braithwaite, 2008) rather than focus on a single business response or a simple dyadic relationship between 'business and society' 'stability and change' or 'regulation and response' we prefer to take a processual approach to the problem. We suggest that multi-layered and increasingly granular 'cycles of regulatory response' will be a useful way to unpick the interaction between firms and their regulatory responsibilities in a way which goes beyond the current literature (e.g. Coglianese and Nash, 2006). Drawing on Van de Ven and Poole (2005) we observed cycles of regulatory responsiveness as the research participants described how their organisations had adapted to the regulations. In both sectors we observed irresolvable tensions between regulatory and commercial priorities. In the retail travel sector we identified three processes which constituted these cycles of responsiveness: recognising, rationalising and refashioning which characterised the phases through which each organisation engaged with the regulations and its implications for them. However we would also like to suggest that these processes operated in a cyclical nature at multiple levels of analysis. In the retail travel sector it was clear that these tensions occurred at multiple analytical levels between a variety of external stakeholders and within the organisation. In this case we saw at least three cycles of responsiveness to the regulations at strategic level and at multiple points in the supply chain. Firms first of all tried to negotiate with the UKBA to determine the form that the eBorders information infrastructure would take and to shape it in a way which supported their business models. Having failed to get the outcome they wanted, they then resolved to keep the regulatory response separate from their commercial operations. However, they eventually chose to capitalise upon it and work it into the customer relationship. We saw further cycles of responsiveness within the organisation as front-line workers and their supervisors integrated eBorders work into their customer management processes. In retail financial services we did not capture firms' initial responses to the regulations as they had been in place for some time. However our study did capture firms' ongoing experiences of living with AML/CTF. Constant operational tensions arose which required adaptation and response in terms of the volumes of regulatory work produced. Financial crime departments had cycles of response in relation to 'evidential spikes'; MLROs pondered the importance of maintaining AML/CTF impetus following highly publicised failures to do so by other banks,[12] customer-facing workers constantly updated their knowledge of what constituted a 'suspicious customer' or 'suspicious transaction' and enacted the regulation in the moment-by-moment interactions with customers. In making this latter point we also acknowledge Sennett's (2006) view that any discourse which implores firms to be flexible and adaptable in response to macro environmental changes is often a double edged sword for those less powerful members of the organisation (see also Roper, James and Higgins, 2005). Therefore it may well be desirable for a firm to make a top level strategic commitment to respond to national security regulations in the way that is required by government. However, the ongoing nature of that commitment is a complex one, featuring response cycles at different organisational layers. Furthermore, multiple responses to different stakeholders are also required, as we saw in the eBorders case in particular. We would also point out that the picture is complicated by the fact that all of our research participants were multiple stakeholders in the regulations. Not only were they members of incumbent organisations, but they were also employees of those organisations, citizens and customers as well.

12 http://www.bbc.co.uk/news/business-20673466 accessed 11th December 2013.

Impact on the sectors: Regulatory fit

Finally we would like to comment on the specific impacts on each sector. Within each sector we have highlighted that there were multi-layered cycles of response to the regulation, which raises the question of whether particular capacities are developed which enable organisations to respond to these kinds of regulation more effectively. We also noted other similarities in response: an increase in staff responsibilisation, the requirements for additional staff training and system adaptations. However these surveillance-based regulations did require different things of organisations. In retail travel, organisations had to gather new data about their customers, integrating it within their customer contact process. In retail financial services they had to analyse consumer data and integrate the process with their customer relationship management processes. We were at pains to point out the potential cross-overs between regulatory responses and commercial activities, particularly around the customer relationship because of primacy of the customer contact point within both regulatory and commercial processes. We illustrated this cross-over using qualitative data which explored different facets of the organisational response around gathering and analysing data; changing job roles for staff; organisational support and systems and responsibilities. These categories were derived from a factor analysis of survey data. Drawing on those survey data further we illustrated how different organisations were at different starting points in their customer relationship management processes. By extension, they would also be at different starting points in their response to AML/CTF because it is so seated within the customer relationship. This certainly seemed to bear out in the correlation tests that were carried out. Significant correlations were found between top management support for CRM and staff empowerment to undertake AML/CTF responsibilities. For us, this raises whether there is a phenomenon of regulatory fit which could be further explored: the extent to which regulatory compliance processes intersect with commercial processes and the extent to which they can support each other. Our data make some preliminary indications that those firms who are more sophisticated in their use of CRM principles and practices experienced some benefits in responses to a set of regulations which relied on firms knowing their customer as part of a commercial strategy. We would like to encourage other researchers to explore this notion. The source of this idea stemmed from surveillance studies, whose concept, 'function creep' or 'expandable mutability' (Norris and Armstrong, 1999) refers to how information gathered for one surveillant purpose can then be used for another purpose. However our question relates to that of organisational capacity rather than use of information, although further consideration is beyond the scope of this book.

Limitations

Before we draw some conclusions, we note the limitations we faced during the course of conducting this piece of research. We would first like to acknowledge, reflexively, the manner in which we have produced the research problematic with which we have engaged. As a group of researchers, we each had different approaches to the problem of understanding what happens to firms when they are mandatorily involved in government national security surveillance programmes. We adopted critical perspectives from surveillance studies, science and technology studies, new media theory, security studies and information infrastructures to try to situate the problem in context. However in order to examine organisational processes and actors, we also used managerialist perspectives from strategic management and marketing combined with more critical work from labour process theory. Combining these perspectives was not easy because of the different interpretations of the phenomena which arose in the data. Strategy and marketing perspectives, for example, addressed staff involvement in AML/CTF as an issue of 'empowerment', whereas labour process perspectives would see it as 'work intensification' or 'responsibilisation'. However as a group of researchers we managed to agree on the main consequences of instantiating a securitised information flow within organisations which implicated the customer relationship and those who managed it. We say more about this process in the Epilogue. Nevertheless we acknowledge that there may be many other ways to dissect this problem, and indeed we have referred to some of them, but our lack of expertise in these areas prevents us from exploring them further in a responsible scholarly way.

Other limitations we experienced surrounded access to participants, particularly for in depth case study work in the financial services sector.

Our key informant data had reached saturation point but we struggled to gain access to financial services organisations for in depth case study work. Instead we relied on a survey which was successfully administered in the sector, although the number of respondents was still small and limited the sophistication of the statistical analysis we could carry out. We hope that we have demonstrated ingenuity in the interpretation and use of our dataset as well as scholarliness.

Conclusion

This research was conceived on a hunch that an organisation's orientation towards its customer may shape its ability to respond to increasing government demands for customer information in an age of pre-emptive security. It was also conceived with a strong belief that the business disciplines could and should be able to make a strong contribution to emergent debates on the surveillance society and the operationalisation of national security. We hope that we have demonstrated the efficacy of those beliefs. The involvement of the private sector in national security arrangements is controversial in the eyes of the general public and the media, and the debate is here to stay. However, whilst advocacy groups, journalists and some branches of the academic community are focused on the rights of citizens under these conditions, our focus has been different. We are concerned with those who are in the middle: employees of all ranks who are mandated to respond while pursuing a commercial agenda. This focus has highlighted how the political economies of private sector organisation have distributed the efforts and rewards associated with involvement in governmental surveillance regimes – and there are no surprises as to who the winners and losers are. Whilst financial services and travel are two important sites, we hope that our insights may inspire others to investigate the communications sector, for whom there is a long road ahead, and for whom the relationship with governments around the world is far from clear cut.

CHAPTER TEN

The out-takes

Reflections on interdisciplinary working

Throughout the research which produced this book we have been aware of our disciplinary backgrounds. They are diverse and at the outset we weren't sure whether we would come together around the business-relevant but also highly critical questions which surround firms' involvement in national security regimes. We wanted to share our experience with all readers so that they might reflect on their own interdisciplinary research practice and perhaps non-business readers would see us business academics in a new light. We decided to sit down and think about the collaborative process. What have we done? How has this project worked? What was involved? How can we understand the workings of the team? Each of us went away and wrote a short reflexive piece on our involvement, which we then coded and reflected on together. Surprisingly, we hit upon Bakhtin's description of the Carnivalesque as an appropriate lens through which to reflect and we generated some interesting insights. Here are some opening remarks which capture the thoughts, feelings or emotions which became enlivened during the research process, as well as the influences that informed and cajoled collaborative interpretations:

> Liz: Previously I have worked closely with other disciplines within business schools, particularly colleagues in marketing, and have always found broad areas of commonality in our approaches. I have not, however, worked with researchers from other areas of social science. Whilst I have found the experience rewarding, I have, and continue to, experience feelings of disorientation and slight discomfort – not because anything they do is wrong, it is just different from how I learnt to approach things. This leaves me feeling that I do not know how to join in their world or work – and what they must make of me.
>
> Kirstie: I am a business academic and I work in a business school. Apparently. And yet my research strength and my research community lies outside that arena. I regularly work with sociologists, geographers, political scientists, international relations specialists, STS people, public admin academics, technologists … the list goes on. They think I'm 'one of them' but I'm not. And yet I never work with business academics. I'm not one of them either …
>
> Maureen: I don't think I was ready for just how different this 'new angle' was going to be! We were a multi-disciplinary team; while some of the other disciplines were ones I had some knowledge of, and felt quite comfortable with (information systems, for example), others were completely new to me (surveillance studies in particular). I can remember reading, and re-reading, papers and book chapters with a mixture of surprise, excitement and bewilderment … and having to look up words that I had never used before and didn't understand …
>
> Keith: Before starting work on [the project] I had been a geographer, my work had centred on consumption and how peoples, cultures and societies buy and sell things. [The project] was a project grounded in Customer Relationship Management, Surveillance and Organisation Studies; none of which I had, at the time, much knowledge of. Surveillance was something I was aware of and was a subject that deeply fascinated me; indeed,

colleagues at my previous institution had been working on many of these issues. CRM and Organisation studies, looking back, did present to me a certain amount of trepidation, here was something that was well outside of my comfort zone and discussions and literatures of business models and statistics puzzled me.

Sally: Working together has had its challenges. We haven't always agreed and sometimes we just didn't 'get' each other at all. This isn't that unusual for me. As a marketer, I'm accustomed to sometimes being cast as from the 'dark side'. You probably know what I'm talking about ... One of those murky types, who subscribes to pestering people with promotions for products they neither need, want, nor can afford. Yet for me these challenges were also a key strength of the collaboration. Our differing perspectives allowed us to take different tangents on the issues we were studying.

Ana: When I joined this project, I already had a multi-disciplinary background. My doctoral thesis – for a degree in Information Systems – had looked at customer profiling issues (not to mention that my first degree was in Economics and my first job in Accounting!). Moreover, I had participated in a Network of Excellence that brought together sociologists, lawyers and computer scientists, among others. So, I was very aware of the distinctive ways of thinking and working of different research communities. And I liked that.

Reflections sensitive to 'doing' research have been illuminative in expanding theoretical appreciations of collaborative work. For instance, one way of thinking reflexively is to write about the 'small' things that have bearings on research. Telephone conversations, driving to research meetings or writing workshops all give unique and often very personal perspectives on doing research (Cann and DeMeulenaere, 2012; Adams and Jones, 2011; Gale et al., 2012). 'Warts and all' approaches to research and the collaborative nature of writing presents an opportunity to firstly pause and think and secondly, to be reflexive on what is being done and how it is being done (MacCormack, 2001). Important also are conceptualisations about how we approach research: effectively 'how we did it'. What, for instance, are the theoretical and philosophical tools we reach for when formulating research questions and are these complicated or simplified in team research?

We argue that there is a need for greater attention on 'doing' qualitative multi-disciplinary research as opposed to modelling the form it takes. We do this by highlighting elements of 'letting go' and 'coming together' when new perspectives and knowledge are mobilised and fortified by a team of researchers. We also introduce the notion of 'carnivalesque collaboration' – the creation of a space that allows researchers to relax their theoretical positions and habits.

Who were we?

The heading is deliberately stated in the past tense because although we originally hailed from different disciplinary backgrounds we are in quite a different place now having completed this work. We are six academics from diverse backgrounds: Kirstie's home discipline is Organisation Theory, Sally is a Marketing academic, Keith is a Human Geographer, prior to working in Information Systems Liz was a Physicist, Ana has worked across Marketing and Information Systems and Maureen specialises in Strategy having been a senior management practitioner in the financial services sector. The following list highlights some of our eclectic theoretical outlooks: marketing strategy, segmentation, customer relationship management (CRM), consumer behaviour, relationship marketing, information systems, information management, organisation theory, the sociology of the body, science and technology studies, regulatory capitalism, new public management, surveillance theory, new media theory, customer profiling, customer management, globalisation and spaces of consumption. Working within these literatures our intention was to gain a deep and detailed understanding of the workings and implications of new regulations in the UK.

The team formed in a relatively straightforward manner; Kirstie was the principal investigator and along with Sally, Maureen and Liz was responsible for writing the original research bid. They were colleagues at the same business school and some of the team had worked together previously, whereas, Keith was recruited to the project and Ana joined in an advisory and participatory role. While team members joined primarily because of their expertise also evident in the team dynamic was a collegial influence (Curry et al., 2012; Massey et al., 2006). This was something that undoubtedly aided our progress and the collaborative way in which we worked. The team was assembled with some consideration of compatibility but it was the skills and knowledge of those involved that guided their participation. Our approach incorporated the strengths of each team member in highlighting important components of regulatory practice within the financial services and travel industries. Each discipline needed to be heard without overly focusing on one discipline and lessening the impact of others – not to mention those disciplines that have not been considered (Hitchings, 2003). Despite our broad antecedents we had a common goal, a focus that positioned our research; for the most part, this focus was the practices and processes that made the regulations work. Our approach considered the surveillant, infrastructural and customer relationship elements that facilitate the process within organisations. The sometimes conflicting, sometimes harmonising theoretical perspectives did eventually fit in the pursuit of our common research goal.

Doing multi-disciplinary research

Theory, Barnes argues (2001), does not have a hold on an exclusive truth, it is messier than that and the 'doing' aspect of theory is equally important. It is an active practice, one with an openness that extends to the debates and discussions that it generates, as well as, its applications and redefinitions in new disciplines and milieu (Lees and Baxter, 2011). Much academic inquiry follows a style of 'theoretical consistency' and approaches often stick to the 'tried and tested' (Newbury, 2011). This is not to dismiss the validity of these approaches, as mono-approaches have fashioned much interesting and informative work. Nevertheless, picking one theory and 'sticking to it' was simply not possible in the domain in which we found ourselves (Taylor, 2007). Indeed Midgley (2011) argues that the choices enacted by the researcher are situated and interlinked by theoretical aims and objectives. The justification in applying particular theories to specific questions is often dependant on the researcher's relationship with 'the wider systems' in which they are positioned. The choice of theory, or even the angle of exploration, that a researcher chooses is relevant to their ontological knowledge and experience (Hardy, Phillips and Clegg, 2001).

Our reflections indicate we were constantly aware of what we brought to the table: literally. The word 'literally' here is used very deliberately to refer to the fact that once a month we brought ourselves – our lived bodies – together around a table to discuss the project and what to do next. This is critical when explaining how the project unfolded because the presence of our lived bodies around a table meant that our respective disciplinary knowledge were brought to bear on the problem at hand through our collective sense-making and discussions enacted co-presently. In writing about these experiences for this book we do not view our knowledge as 'out there' but as physically embodied and produced as part of a material interaction in a particular space and time. Furthermore our different positions in relation to the material did not just stem from our disciplinary backgrounds, they were produced by the surveillant phenomenon we were investigating. The phenomenon required an articulation and integration of its many facets. Barad (2003: 815) refers to this as 'intra-action':

> It is through specific agential intra-actions that the boundaries and properties of the "components" of phenomena become determinate and that particular embodied concepts become meaningful.

Within these intra-actions and boundary negotiation lies an inherent tension. The research process needs to maintain certain conventions e.g. producing new knowledge which is publishable, contains valuable insights and satisfies the funders. And yet it must also overcome the conventionalities and constraints of the researchers' disciplinary foundations to get to that new knowledge. It is precisely these aspects of academic practice which bind the researchers together but at several moments has the potential to divide them. As Star and Griesemer's (1989) argue, there needs to be a plasticity in the practices, materialities and embodied sensibilities which combine during the collaborative process. A 'standard' is needed to theorise, without this we cannot distinguish one understanding from another, and herein lies the tension to collaborative work (Metro-Roland, 2010; Deleuze and Guattari, 1994). One person believes in doing it one way and the other in doing in a different way and somebody has to 'let go'. The standard is the anchor, the collaboration the plasticity. However, what of the movement of anchor to plastic, and how this in turn encourages new knowledge? What kinds of embodied sensibilities were enacted? There are two elements to this process upon which we explicitly reflect: Letting go of our disciplines and creating the space that enabled the collaboration to emerge.

The letting go

Being involved in this project entailed some uncomfortable and sometimes painful moments of departure; true to form, feelings of anxiety and insecurity ensued. Moving away from our theoretical histories and perspectives, as well as learning new approaches and enjoying new hybrid theoretical outlooks was eventful. Each researcher was required to adopt a degree of plasticity in order to adapt to the new working arrangements. This was problematic in that it distanced us from our disciplinary roots and compromised the degree to which we 'fit' with one another. We were then trying to create new, hybrid and cohesiveness ways in which to be 'rooted', adapting and acknowledging others' positions as we did so.

The picture was far from clear. Kirstie questions whether her colleagues had the same motivations as she did, using emotive language and identifying emotion as a powerful driver (Stürmer and Simon, 2009) behind her involvement in the project:

> Kirstie: I don't know that they feel the same degree of anger as I do at the unfairness of social sorting and the way it concentrates power in untransparent ways. I hope that they caught a few glimpses of that and that I made the critique available in a way that was useful to them.

Indeed she also comments on the constant sense of plasticity inherent in the movement between disciplinary and multi-disciplinary worlds. She highlights the ambivalence and struggle she feels in how she identifies herself and in how she approaches her work.

> Kirstie: [I have experienced] years of teaching related boredom, intellectual isolation and the rejection of my work from what I thought were journals and communities directly representing my home discipline 'organisation studies'. This has also led to a sense that I have been 'moonlighting' in the surveillance studies world – a community of loosely affiliated misfits ... It seems detached from my everyday life. It feels unreal and something that my everyday colleagues cannot relate to. I'm working 'in-between'.

Keith also expresses the discomfort he felt at the start of the project, having moved from a geography department to a business school, and then to a multi-disciplinary surveillance project:

> Keith: some things are done differently at a business school – for instances, it is necessary to engage with organisations and business and all that this entails (e.g. established links with banks or small business; suited visitors to the department) and academic staff often have consultancy positions with non-academic organisations, these were things that I had not experienced in geography departments … other things were different, advertisements for invited speakers to the school spoke of mysterious and unusual titles 'international management practice' 'intangible assets', 'managerial innovations' – I had heard these terminologies before but had no idea what they meant.

The juxtaposition for us is that we were limited by our antecedents but unbounded in our collaborative potential. Maureen comments that, in the end, she has a sense that we managed each to situate ourselves in the new multi-disciplinary context, but that she cannot really be sure:

> Maureen: Given that the literature stresses how difficult it can be to communicate across disciplines or knowledge domains, I think we have (probably, mostly!) managed to do this well? … It would be interesting to track how some of our … draft papers … evolved over time; and also whether our level of consensus has changed (do we agree with each other more, or less, than we did at the beginning of the project about issues such as surveillance practice or CRM practice, for example? Or doesn't it matter?) And I really like the idea of us using these objects to create knowledge in different domains, as well as at the intersection of those domains.

Furthermore the plasticity required enabled a reflective distance to emerge. There is an unacknowledged imperative of fit, that haunts much academic work and as evidenced in our reflections this it seems is the biggest obstacle to plasticity. As soon as one becomes aware of one's disciplinary limitations, one questions the limitations of one's own contribution. Liz comments on how she became aware of her disciplinary vernacular and its limitations during the project:

> Liz: The two aspects that strike me when considering working in the project – both on empirical aspects and theorising – are: both my own and the IS field's (overly?) rational and pragmatic approach compared to other disciplines, the need to recognise and incorporate both detail and larger scale concepts and finally, perhaps most strikingly, both my own and the IS field's lack of critical engagement in our studies of emergent IS phenomena.

As the project advanced, we contended with some trepidation as we moved into the unknown while maintaining our disciplinary identities as the anchor. As results materialised and we engaged with the world of publishing, it emerged that too much plasticity resulted in a lack of focus. Ana continues on tensions that have arisen when submitting the project's work for publication, as the conventional, in Star and Griesemer's (1989) terms, begins to reassert itself:

> Ana: We have now reached the stage to focus on disseminating the research outputs. The project's multi-disciplinary approach was welcomed by the sponsor and the findings well received among the practitioner community. However, it has been a challenge to get our work accepted in prestigious academic outlets. The wide body of literature is deemed a weakness. Instead, we are asked to narrow the scope, while exploring discipline-specific topics in more depth. The terminology becomes another issue. We need to adapt terms that are familiar to the audience of the target journal, even if they fail to capture the full breadth of what we are talking about. And the journey between the macro, meso and micro levels is difficult to capture in the limited words available for the traditional journal paper. It seems that, as far as publications are concerned, 'multi-disciplinary' is, again, seen as lacking focus.

There is a warm appreciation here of difference, yet underpinning the reflection is how it 'lacks focus' or 'things are done differently'. Despite our reservations and trepidations what we can view in our reflections is an undercurrent of desire to collaborate. More effusive is a tentative acceptance of moving into the unknown, or working 'in-between'. All of us see collaborative work positively, but we qualify it with our anxieties, striving to retain a connection with our disciplinary 'homes'. As Barnes (2001) has argued theorising is an activity and as such, the trepidations that come with attempting something for the first time are part of the theoretical process. We associate with established disciplines, theories and practices and moving away and between those enacts new ways of critiquing the 'truth', but also exposes researchers to critique from within their 'standard' approach. We are conscious of both.

The carnivalesque collaboration

In thinking-through our collective 'letting go' we consider the manner in which we merged in our research focus and use Bakhtin's concept of carnivalesque to depict what happened. For Bakhtin, carnival is an allegory for transformation, where the world is turned upside down (White, 1993). Bakhtin sets his argument in the world of Rabelais, a 16th century French renaissance writer, and challenges the 'sanitized' bourgeois version of the 'self'. As Hall (1993: 7) states, 'for Bakhtin, this upturning of the symbolic order gives access to the realm of the popular – the "below", the "underworld", and the "march of the uncrowned gods"'. The carnival offers an ambivalent perspective as carnival goers participate in events mixed between reality and make-believe. For Bakhtin there is a suspension that enables *new modes of interrelations*, where those who do not normally mingle or converse do so. *Eccentricity* is prevalent when behaviours viewed as alternative or different are celebrated for their performative zeal. In addition, a *carnivalistic mésallinances* is encouraged where formalities and distinctions are ignored for the duration of the carnival and lastly there is room for *Profanation* as obscenities and ridicule of those of higher social order is tolerated (see Sandner, 2004). Carnival offers difference in much the same way as the collaborative nature of our work. What is turned-upside is our academic fit – the existing academic world from which we came. There was a collective letting go and then coming together in how we merged within the project. The space we created was a space which was away from the everyday, perhaps our equivalent of the festival, where existing ways of working were turned on their heads. This we contend speaks to Bahtkin's *carnivalistic senses of the world*.

The sensuousness gaiety, talk of defecation and general billingsgate of the fairs, feast and festivals in the Middle Ages were for Bahktin unlike the seriousness of ecclesiastical and feudal life. Folk-culture besmirched high-culture and encouraged temporal inversions – the 'fool' become king for the day (Bakhtin, 1984: 10). However, the 'peculiar culture', its element of laughter and the release from the everyday are performative rather than political (Ravenscroft and Gilchrist, 2009). There is always an assumption that hierarchy and order will be returned. After all, the festivals are in many ways structured, they follow dates in the calendar and follow etiquettes and traditions and possibly most importantly they are sanctioned by the religious or feudal powers of authority. Bahktin (1984: 154) states:

> the Renaissance had its own territory and its own particular time, the time of fairs and feasts. This territory, as we have said was a peculiar second world within the medieval order and was ruled by a special type of relationship.

When we were reflecting on our experiences and Bakhtin's perspective, we didn't see the carnival as a space of political upheaval, rather we saw it as a place of expression, sanctity, reprieve and suspension that encourages multi-disciplinary development. There is still politics here, but one overridden by a 'letting go' and immersing oneself in the polyvocality of a research team. Traces of sensitivities and hierarchies remain, much like the structure of Bahktin's carnival, we are aware we will return to our academic positions when outside the room, but there is an awareness of collaboration and what that demands. As Sally suggests:

> Sally: It's fair to say that there have been moments during the project when we have had to walk on egg shells, being courteous and sensitive to one another's needs and perspectives. Funnily enough, I found this behaviour to be in itself, hugely constructive. On other projects I work regularly with my husband, who is also an academic. When you know someone so well, it is too easy (although not always desirable), to dispense with polite, co-worker civilities. The danger in so doing is that we can forget to listen, and may fail to hear or respect the alternative point of view.

While academic life is hierarchical, political and competitive in its nature and the team members ranged in experience and positions of esteem, there was a levelling within the dynamic of the team (Curry et al., 2012). While seniority can never be truly left outside the door, nor indeed the different and strange ways others do things, there is a temporal suspension of the everyday and the collaborative spirit it induces channels ontological awakenings.

Infused within our collaborative carnivalesque was a strong sense of engaging with words and ideas that one perhaps would not have done in a disciplinary context. What one can say and do is a matter of disciplinary power and in our case was circumscribed by our academic disciplines. While others have drawn on the transgressive nature of the carnival and its relation to deviant or competitive behaviours (Clisby, 2012; Ravenscroft and Gilchrist, 2009), we emphasise the reflexive potential when daily pressures are suspended. 'Extraterritoriality' or 'place-beyond-place', as Bakhtin (1984) would have it, becomes place to collaborate. Having a collaborative space has throughout the project been important in how we have worked together. Each of us was 'the only expert in the room' in relation to their respective disciplinary knowledge. We each brought with us a knowledge with which others on the team were less familiar. As a result team members at all career stages enjoyed status and the political or competitive element of colleagues working in the same field was alleviated. All of this nurtured an escape or evasion from the usual 'official order and official ideology' (Bakhtin, 1984: 154), as well as a distancing between the carnival and 'real world' elements of academia (Stallybrass and White, 1986). We referred to it as 'the research sanctuary' and frequently breathed a sigh of relief as we closed its door, shutting out the rest of the school and beginning our deliberations.

The coming together

Although we began with anxieties of working on a 'risky' project, the central terrain of doing theory in a carnivalesque space is that understandings of otherness and alterity began to surface (Folch-Serra, 1990). This in turn fuelled our thinking as we took on board others' expertise and perspectives. We searched for theoretical frameworks in a number of different disciplines that might explain our observations, but we drew a blank. The phenomenon we were interested in explaining did not feature neatly anywhere so we had to find our own way of describing it. Eventually our discussions identified a concept to which each of us could relate. We settled on the concept of a 'securitised information flow' as the unit of analysis and about which theory could be generated. The 'securitised information flow' described the phenomenon of customer data sharing with government in a way that incorporated all of our disciplinary interests as well as describing the surveillant phenomena at hand. The word 'securitised' was helpful as it distinguished the type of information which was flowing for commercial purposes, such as CRM related or employee performance information. We saw 'information flow' as the conduit which linked the organisation, its customers, workers and information systems, with government. Team members were comfortable with how organisational resources – and the concepts they knew about the best – became 'aligned' with and invested in the continuation of the flow. It was viewed as something which translated customers as: i) customers whose activities generated data for strategic commercial purposes (Sally, Maureen, Ana); ii) into potential threats to national security (Kirstie); iii) positioned the organisation and its members as an important intermediary (Keith). In addition, employees brought the worlds of business and regulation together (Kirstie) as they plugged infrastructural gaps between customers and information infrastructures to ensure that data flowed (Liz).

Echoing Barnes' (2001) observations that theorisation is receptive to diversity and as we have reflected, differing approaches and values produced insightful knowledges. Kirstie describes the feeling of convergence which emerged:

> Kirstie: As the project progressed we also found that we had a lot in common as we pragmatically solved research design and execution issues. We created new ways of describing the phenomena we were observing, such as 'remediation work' and 'aligned elements of a securitised information flow' which integrated our perspectives. We also opened up publication avenues through our individual connections and expertise. We shared the frustrations of it too. The work itself is now rooted in the business literatures which we shall make available to everyone else through publication ...

Working with others of a different theoretical persuasion, has forced us to think about our own position first, then that of other team members and lastly to the perspectives of our audiences. Each of these persuasions has been influential in the practicalities and situated practices of engaging with the research project. A balance of sorts is struck in how we as researchers rationalise our approaches and inputs. We play to our 'strengths' but we are, particularly as academics, aware of our 'weaknesses' – especially when exposed to blind review. Ana writes:

> Ana: The other members of the team were very constructive in their approach to the project. No one ever tried to impose their discipline as the best one to study the problem, or their methodology as the right one. Though, equally, no one took the other's explanation or vocabulary as given, and we often spent considerable time discussing the background to a particular concept or framework, and agreeing terminology. It seems to me that, to work in a multi-disciplinary team, it is necessary to both have confidence in what you know and willingness to question your long established assumptions.

Ultimately the social practices of debate, encouragement and camaraderie have driven the intellectual impetus inherent in our project (Hirsch, 1967; Dilthey, 1996). As Ana points out, we begin to see what we have done, how we have diversified and to where we have moved theoretically. Liz continues:

> Liz: Whilst I have never thought of myself as embracing a positivist epistemology ... It would appear that an individual considered as an interpretivist in the IS field, may appear as hard-line positivist by organisational studies standards ... This difference in latent epistemologies may have been a source of potential incompatibility but I do not perceive it has been. Rather, I think the team has identified and drawn on the benefits of these different approaches, combining the detail of reality that is often characterises pragmatic and grounded IS studies with the more wide reaching generalities of organisational studies.

The intricacies of converging disciplines have reversed some of Liz's theoretical positionality (see England, 1994) and working from a reflexive imperative, and one grounded in the skills and experiences of a diverse research team does question and expand how theory is used critically.

Final reflections

Doing research and engaging with multi-disciplinary is as Bourdieu and Wacquant (1992) suggest hindered by 'abstraction' or the role of translating the highly complex into the readable and scientific. Theory of course offers some help. In order to learn 'from' and not 'about' requires a more substantive appreciation of what is being observed or experienced, and within this is a positional recognition of where the reader is and where the researcher is (Laurier, 2001). As Garfinkel (1974: 11) states, 'common sense' and knowledge of ordinary affairs are 'organised enterprises, where that knowledge is treated by us as part of the same setting that also makes it orderable'. Just as practical action and practical reasoning embody observable detail (see Garfinkel, 1996; Laurier and Philo, 2006), attention should be applied to the contexts that surround and enable new knowledge accumulation. Reflecting on our reflections, as we have all done when reading the drafts of this book, we do remain a little puzzled as to when we 'let go' and 'how did we know we let go'? As Sally questions in her account:

> Sally: And finally, I have a further and perhaps contradictory thought about cross-disciplinary working. In spite of all that I have said about the positives of such working, and even though this sharing of differing ideas and insights allows us to reflect and grow, I have also found it important to hold fast to my own research identity. It is mine after all, and I, like my fellow team members, have worked hard to develop it and have earned the right to own it.

Possibly, like Garkinkel (1974), we struggle to acknowledge what was taken-for-granted and toil to articulate when and where we 'let go'. Or, possibly much like the world of research, and indeed Bahktin's world of laughter and Star and Greisemer's (1989) plasticity, there is an in-betweeness to our areas of expertise (Levinas, 1986). We as researchers are structured by our disciplines, research methods, research teams and research audiences and we rightly celebrate this and are protective of it. When we research using an unfamiliar method or writing about a new theory we are not irrevocably changed as researchers, instead we have just expanded our research perspectives and experiences. What is essential is a willingness to work in this fashion and possibly this is why the dynamic of the research team was productive. Maybe the strength of the team grew because our personalities were compatible – we all, without acknowledging it directly, accepted and welcomed the perspectives of others and were committed to work with them. We let go as soon as we agreed to join the team and each time we entered a research meeting we performed further letting gos.

Along the way we have found characteristics of plasticity and carnival that have supported those letting go and fortifying moments. Because of this the research has been taken and directed on numerous routes. In each and every instance it was the collaborative process that guides this. However, writing our theoretical journey is less than straightforward. We have gone some way to exposing how we, as six researchers, felt in how we participated on this project and how our findings were moulded firstly, by our trepidations and our comfort with what we know; secondly, our comfort was disrupted by listening to and working with others who have differing histories and perspectives. And lastly, the work progressed through the reflective space generated by the team dynamic. The experience of working on the project, all of us agree, was positive in the extreme, yet there is no guarantee that future collaborative work will be the same. To some degree we all still hold reservations and we will all ultimately return to our disciplines of choice, and to our 'fit'. Surveillance studies has an extra-territorial quality within the academy, in that it is 'outside' or 'away' from the day to day practices of discipline-based research. Each of us might reflect on the quality of collaborative spaces within surveillance studies as well as the importance of embodied knowledges situated within ourselves as we grapple with new and emergent phenomena without our home disciplines.

Our experience also gave the collaboration around questions of surveillance – which was a new idea to engage with for all but one of us – an attractive subversive quality. For the majority of colleagues on the project a temporary subversion which was as empowering as it was daunting: Carnivalesque in a small scale, north European, sort of way. There remains the pressing acknowledgement that collaborative work brings new knowledge and meaning together. Ultimately, much like Gadamer (1992) what underscores our contemplations is an understanding of what has made this possible. In developing new found knowledge, plasticity, embodied sensibilities and, our intention has been to rationalise our experiences and to promote the theoretical expansion of thinking about doing multi-disciplinary qualitative research. There is good and bad in this. We have had difficulties publishing this work in journals which have strong disciplinary roots, even when they purport not to (!), but we feel our contributions to the literature are robust as a result of thinking differently. Our reflections made us think about our fit in a multi-disciplinary team, both issues are important and need careful consideration, how else can we begin to understand how we understand?

References

Ackermann, F. and Eden, C. (2011). Strategic management of stakeholders: Theory and practice. *Long Range Planning*, 44(3), 179–196.

Adams, T. E. and Jones, S. H. (2011). Telling stories: Reflexivity, queer theory, and autoethnography. *Cultural Studies/Critical Methodologies*, 11(2), 108–116.

Adey, P. (2009). Facing airport security: affect, biopolitics, and the pre-emptive securitisation of the mobile body. *Environment and Planning D: Society and Space*, 27(2), 274–295.

Adey, P., Brayer, L., Masson, D., Murphy, P., Simpson, P. and Tixier, N. (2013). 'Pour votre tranquillité': Ambiance, atmosphere, and surveillance. *Geoforum*, 49, 299–309.

Aglietta, M. (2000). *A Theory of Capitalist Regulation: The US Experience* (Vol. 28). London: Verso.

Amicelle, A. (2011). Towards a 'new' political economy of financial surveillance. *Security Dialogue*, 42(2), 161–178.

Amicelle, A. and Favarel-Garrigues, G. (2012). Financial surveillance: Who cares? *Journal of Cultural Economy*, 5(1), 105–124.

Amoore, L. and de Goede, M. (2008). Transactions after 9/11: The banal face of the preemptive strike. *Transactions – Institute of British Geographers*, 33(2), 173–185.

Amoore, L. (2009). Algorithmic war: Everyday geographies of the war on terror. *Antipode: A Radical Journal of Geography*, 41(1), 49–69.

Andrejevic, M. (2009). *iSpy: Surveillance and Power in the Interactive Era*. Lawrence: University Press of Kansas.

Aradau, C., Lobo-Guerrero, L. and van Munster, R. (eds.) (2008). Special issue on security, technologies of risk, and the political. *Security Dialogue*, 39(2–3), 147–357.

Archer, M. S. (1995). *Realist social theory: The morphogenic approach*. Cambridge: Cambridge University Press.

Arnaboldi, M. and Spiller, N. (2011). Actor-network theory and stakeholder collaboration: The case of Cultural Districts. *Tourism Management*, 32(3), 641–654.

Assets Recovery Agency (2003). *Annual Report 2002/2003*. London: Assets Recovery Agency.

Aufhauser, D. D. (2003). Terrorist financing: Foxes run to ground. *Journal of Money Laundering Control*, 6(4), 301–305.

Backhouse, J., Demetis, D., Dye, R., Canhoto, A. and Nardo, M. (2005). *Spotlight: New approaches to fighting money-laundering* (4 volumes), AGIS programme JAI/2004/AGIS/182.

Bailur, S. (2006). Using stakeholder theory to analyze telecentre projects. *Information Technologies and International Development*, 3(3), 61–80.

Baker, D. P. and Pattison, J. (2012). The principled case for employing private military and security companies in interventions for human rights Purposes. *Journal of Applied Philosophy*, 29(1), 1–18.

Bakhtin, M. (1984). *Rabelais and his World*. Bloomington: Indiana University Press.

Ball, K. (2002). Elements of surveillance: A new framework and future directions. *Information, Communication and Society*, 5(4), 573–590.

Ball, K. (2009). Exposure: Exploring the subject of surveillance *Information, Communication & Society*, 12(5), 639–657.

Ball, K., Haggerty, K. and Lyon, D. (eds.) (2012). *Routledge Handbook of Surveillance Studies*. London: Routledge, 292–300.

Ball, K. and Snider, L. (eds.) (2013). *The Surveillance Industrial Complex: Towards a Political Economy of Surveillance*. London: Routledge.

Ball, K., Canhoto, A. I., Daniel, E., Dibb, S., Meadows, M. and Spiller, K. (2013). Working on the Edge: Remediation work in the UK Retail Travel Sector. *Work, Employment and Society*, 28(2), 305–322.

Barad, K. (2003). Posthumanist performativity: Toward an understanding of how matter comes to matter. *Signs*, 28(3), 801–831.

Barnes, T. (2001). Retheorizing economic geography from the quantitative revolution to the 'cultural turn'. *Annals of the Association of American Geographers*, 91(3), 546–565.

Basel Committee on Banking Supervision (2001). *Customer Due Diligence for Banks*. Basel: Basel Committee on Banking Supervision.

BBC (2012). Brian Wheeler, *The Truth Behind UK Migration Figures*, available from http://www.bbc.co.uk/news/uk-politics-19646459, accessed March 2013.

BBC (2014). *Moneybox*, 4 January. http://news.bbc.co.uk/1/shared/spl/hi/programmes/money_box/transcripts/money_box_04_jan_14.pdf, accessed March 2014.

BBC Persian (2012). Bahman Kalbasi, *TD Bank: Iranian-Canadians Caught Up in Sanctions Row* http://www.bbc.co.uk/news/world-us-canada-18803129, accessed March 2014.

BBC (2012). *Draft Communications Data Bill to be Redrafted – No 10* http://www.bbc.co.uk/news/uk-politics-20676284, accessed December 2012.

Beniger, J. R. (1986). *The Control Revolution: Technological and Economic Origins of the Information Society*. Boston: Harvard University Press.

Benington, J. (2007). From private choice to public value? In: Bennington, J. and Moore, M. (eds.), *In Search Of Public Value – Beyond Private Choice*. London: Palgrave.

Benjamin, R. I. and Levinson, E. (1993). A framework for managing IT-enabled change. *Sloan Management Review*, 34(4), 23–33.

Bennear, L. S. (2006). Evaluating management based regulation. In: Coglianese, C. and Nash, J. (eds.), *Leveraging the Private Sector*. Washington: RFF Press, 55–86.

Bennett, C. J. (2001). Cookies, web bugs, webcams and cue cats: Patterns of surveillance on the world wide web. *Ethics and Information Technology*, 3(3), 197–210.

Bennett, C. J. and Haggerty, K. (eds.) (2011). *The Security Games: Surveillance and Control at Mega Events*, London: Routledge.

Bhatt G., Emdad, A., Roberts, N. and Grover, V. (2010). Building and leveraging information in dynamic environments: The role of IT infrastructure flexibility as enabler of organizational responsiveness and competitive advantage. *Information & Management*, 47(5/6), 341–349.

Bigus, O. E. (1972). The milkman and his customer: A cultivated relationship. *Journal of Contemporary Ethnography*, 1(2), 131–165.

Bohling, T., Bowman, D., Lavalle, S., Mittal, V., Narayandas, D., Ramani, G. and Varadarajan, R. (2006). CRM implementation: Effectiveness issues and insights. *Journal of Service Research*, 9(2), 184–194.

Bolter, J. and Grusin, R. (2000). *Remediation: Understanding New Media*. Cambridge: MIT Press.

Boonstra, A., Boddy, D. and Bell, S. (2008). Stakeholder management in IOS projects: Analysis of an attempt to implement an electronic patient file. *European Journal of Information Systems*, 17(2), 100–111.

Boudreau, M. and Robey, D. (2005). Enacting integrated information technology: A human agency perspective. *Organization Science*, 16(1), 3–18.

Boulding, W., Staelin, W., Ehret, M. and Johnston, W. J. (2005). A customer relationship management roadmap: What is known, potential pitfalls, and where to go. *Journal of Marketing*, 69(4), 155–166.

Bourdieu, P. and Wacquant, L. (1992). *An Invitation To Reflexive Sociology*, Chicago: University of Chicago Press.

Bowker, G. C. and Star, S. L. (1999). *Sorting Things Out: Classification and its Consequences.* Cambridge: MIT Press.

Boyatzis, R. E. (1998). *Transforming Qualitative Information: Thematic Analysis and Code Development.* London: Sage.

Braa, J., Hanseth, O., Heywood, A., Mohammed, W. and Shaw, V. (2007). Developing health information systems in developing countries: The flexible standards strategy. *MIS Quarterly,* 31(2), 381–402.

Braithwaite, J. (2008). *Regulatory Capitalism: How It Works, Ideas For Making It Work Better.* Cheltenham: Edward Elgar.

Bryman, A. (2004). *Research Methods and Organization Studies* (Vol. 20). London: Routledge.

Burawoy, M. (1985). *The Politics of Production: Factory Regimes Under Capitalism and Socialism* London: Verso.

Canhoto, A. (2008). Barriers to segmentation implementation in money laundering detection. *Marketing Review,* 8(2), 163–181.

Canhoto, A. and Backhouse, J. (2005). Tracing the identity of a money launderer. In: Nabeth, T. (ed.), *Fidis D2.2: set of use case and scenarios.* Fontainebleau: Insead, 35–38.

Canhoto, A. and Backhouse, J. (2007). Profiling under conditions of ambiguity: An application in the financial services industry. *Journal of Retailing and Consumer Services,* 14(6), 408–419.

Cann, C. N. and DeMeulenaere, E. J. (2012), Critical co-constructed autoethnography. *Cultural Studies <=> Critical Methodologies,* 12(2), 146–158.

Capon, N. (1982). Credit scoring systems: A critical analysis. *Journal of Marketing,* 46(2), 82–91.

Chia, R. (2004). Strategy-as-practice: Reflections on the research agenda. *European Management Review,* 1(1), 29–34.

Chuang, S. and Lin, H. (2013). The roles of infrastructure capability and customer orientation in enhancing customer-information quality in CRM systems: Empirical evidence from Taiwan. *International Journal of Information Management,* 33(2), 271– 281.

Ciborra, C. (ed.) (2000). *From Control to Drift: The Dynamics of Corporate Information Infrastructures.* Oxford: Oxford University Press.

Clisby, S. (2012). Summer sex: Youth, desire and the carnivalesque at the English seaside. In: Donnan, H. and Magowan, F. (eds.), *Transgressive Sex: Subversion and Control in Erotic Encounters.* Oxford: Berghan, 47–68.

Coglianese, C. and Nash, J. (2006). *Leveraging the Private Sector.* Washington: RFF Press.

Crang, P. (1994). It's showtime: On the workplace geographies of display in a restaurant in southeast England. *Environment and Planning D: Society and Space,* 12(6), 675–704.

Curry, L. A., O'Cathain, A., Clark, V. L. P., Aroni, R., Fetters, M. and Berg, D. (2012). The role of group dynamics in mixed methods health sciences research teams. *Journal of Mixed Methods Research,* 6(1), 5–20.

de Goede, M. (2008).The politics of preemption and the war on terror in Europe. *European Journal of International Relations,* 14(1), 161–185.

de Goede, M. (2012). *Speculative security: The politics of pursuing terrorist monies.* Minneapolis: University of Minnesota Press.

Deleuze, G. and Guattari, F. (1987). *A Thousand Plateaus,* Trans. Brian Massumi. Minneapolis: The University of Minnesota Press.

Deleuze, G. and Guattari, F. (1994). *What is Philosophy?* London: Verso.

De Luca, L. M. and Atuahene-Gima, K. (2007). Market knowledge dimensions and cross-functional collaboration: Examining the different routes to product innovation performance. *Journal of Marketing,* 71(1), 95–112.

DeRosa, M. (2004). *Data Mining and Data Analysis for Counterterrorism.* Washington DC: Center for Strategic and International Studies, 6–8.

Dibb, S., Ball, K., Canhoto, A., Daniel, E., Meadows, M. and Spiller, K. (2014). Taking Responsibility for Border Security: Commercial Interests in the face of e-Borders. *Tourism Management,* 42(1), 50–61.

Dibb, S. and Meadows, M. (2004). Relationship Marketing and CRM: a financial services case study. *Journal of Strategic Marketing,* 12(2), 111–125.

Dibbern, J., Winkler, J. and Heinzl, A. (2008). Explaining variations in client extra costs between software projects offshored to India. *MIS Quarterly,* 32(2), 333–366.

Dilthey, W. (1996). Hermeneutics and its history. In: Makkreel, R. and Rodi, F. (eds.), *Wilhelm Dilthey: Selected Works: Vol. 4: Hermeneutics and the Study of History.* Princeton: Princeton University Press, 33–258.

Donaghy, M. (2002). Monetary privacy in the information economy. *International Sociology,* 17(1), 113–133.

Donaldson, T. and Preston, L. (1995). The stakeholder theory of the corporation: Concepts, evidence and implications. *Academy of Management Review,* 20(1), 65–91.

Due, C., Connellan, K. and Riggs, D. W. (2012). Surveillance, security and violence in a mental health ward: An ethnographic case-study of an Australian purpose-built unit. *Surveillance & Society,* 10, 292–302.

Eden, C. and Ackermann, F. (1998). *Making Strategy: The Journey of Strategic Management.* Sage: London.

Edwards, P. N., Bowker, G. C., Jackson, S. J. and Williams, R. (2009). Introduction: An agenda for infrastructure studies. *Journal of the Association for Information Systems*, 10(5), 364–374.

England, K. (1994). Getting personal: Reflexivity, positionality, and feminist research. *Professional Geographer*, 46(1), 80–89.

Ericson, R. and Doyle, A. (2006). The institutionalization of deceptive sales in life insurance: Five sources of moral risk. *British Journal of Criminology*, 46(6), 993–1010.

Ericson, R. and Haggerty, K. (1997). *Policing the Risk Society.* Toronto: University of Toronto Press.

Fassin, Y. (2009). The stakeholder model refined. *Journal of Business Ethics*, 84(1), 113–135.

FATF (2010). *Global Money Laundering & Terrorist Financing Threat Assessment – A view of how and why criminals and terrorists abuse finances, the effect of this abuse and the steps to mitigate these threats.* Paris: Financial Action Task Force.

FATF (2010). *Financial Action Task Force Annual Report 2010–2011.* http://www.fatf-gafi.org/topics/fatfgeneral/documents/fatfannualreport2010-2011, accessed 5 January 2011.

Finn, R. L. and Wright, D. (2012). Unmanned aircraft systems: Surveillance, ethics and privacy in civil applications. *Computer Law & Security Review*, 28(2), 184–194.

Finnegan, D. J. and Currie, W. L. (2010). A multi-layered approach to CRM implementation: An integration perspective. *European Management Journal*, 28(2), 153–167.

Flecker, J. and Meil, P. (2010). Organisational restructuring and emerging service value chains: Implications for work and employment. *Work, Employment and Society*, 24(4), 680–698.

Folch-Serra, M. (1990). Place, voice, space: Mikhail Bakhtin's dialogical landscape. *Environment and Planning D: Society and Space*, 8(3), 255–274.

Ford, R. (2010). Electronic border-check firm is sacked; Home Office axes contract over year-long delay. *The Times*, 23 July.

Foucault, M. (1977). *Discipline and Punish: The Birth of the Prison.* Harmondsworth: Penguin.

Freeman, R. E. (1984). *Strategic Management: A Stakeholder Approach.* Boston: Pitman.

Friedman, A. L. and Miles, S. (2002). Developing stakeholder theory. *Journal of Management Studies*, 39(1), 1–21.

FSA (2003). *Reducing Money Laundering Risk: Know Your Customer and Anti-Money Laundering Monitoring.* London: Financial Services Authority.

Fussey, P. and Coaffee, J. (2011). Olympic rings of steel: Constructing security for 2012 and beyond. In: Bennett, C. J. and Haggerty, K. (eds.), *The Security Games: Surveillance and Control at Mega Events.* London: Routledge, 36–54.

Gad, C. and Lauritsen, P. (2009). Situated surveillance: An ethnographic study of fisheries inspection in Denmark. *Surveillance & Society,* 7(1), 49–57.

Gadamer, H. (1992). *Truth and Method.* New York: Crossroads.

Gal, U., Lyytinen, K. and Yoo, Y. (2008). The dynamics of IT boundary objects, information infrastructures, and organizational identities: The introduction of 3D modelling technologies into the architecture, engineering, and construction industry. *European Journal of Information Systems,* 17(3), 290–304.

Gale, K., Martin, V., Sakellariadis, A., Speedy, J. and Spry, T. (2012). Collaborative writing in real time. *Cultural Studies <=> Critical Methodologies,* 12(5), 401–407.

Gandy Jr., O. H. (1993). *The Panoptic Sort: A Political Economy of Personal Information.* Boulder: Westview Press.

Gandy Jr., O. H. (2009). *Coming to Terms with Chance: Engaging Rational Discrimination and Cumulative Disadvantage.* Aldershot: Ashgate.

Gandy Jr., O. H. (2010). Engaging rational discrimination: Exploring reasons for placing regulatory constraints on decision support systems. *Ethics and Information Technology,* 12(1), 29–42.

Garfinkel, H. (1974). The origins of the term ethnomethodology. In: Turner, R. (ed.), *Ethnomethodology.* Harmondsworth: Penguin Books, 13–18.

Garfinkel, H. (1996). Ethnomethodology's program. *Social Psychology Quarterly,* 59(1), 5–21.

Gates, K. (2012). The globalization of homeland security. In: Ball, K., Haggerty, K. and Lyon, D. (eds.), *Routledge Handbook of Surveillance Studies.* London: Routledge, 292–300.

Geiger, S. and Turley, D. (2005). Socializing behaviors in business-to-business selling: an exploratory study from the Republic of Ireland. *Industrial Marketing Management,* 34(3), 263–273.

Genosko, G. and Thompson, S. (2009). *Punched Drunk: Alcohol, Surveillance and the LCBO 1927–1975.* Black Point: Fernwood Publishing Company.

German, M. and Stanley, J. (2008). Fusion center update. *American Civil Liberties Union,* https://www.aclu.org/sites/default/files/pdfs/privacy/fusioncenter_20071212.pdf, accessed 11 April 2014.

Gerth, H. H. and Wright Mills, C. (1970). *Character and Social Structure: Psychology of Social Institutions.* New York: Harcourt, Brace and World Inc.

Gill, M. and Taylor, G. (2003). Can information technology help in the search for money laundering? The view of financial companies. *Crime Prevention and Community Safety,* 5(1), 39–47.

Gill, M. and Taylor, G. (2004). Preventing money laundering or obstructing business? Financial companies' perspectives on 'Know Your Customer' procedures. *British Journal of Criminology,* 44(4), 582–594.

Gilling, D. and Schuller, N. (2007). No escape from the iron cage? Governmental discourse in new labour's community safety policy. *Crime Prevention & Community Safety,* 9(4), 229–251.

Gilliom, J. (2001). *Overseers of the Poor: Surveillance, Resistance and the Limits of Privacy.* Chicago: Chicago Series in Law and Society.

Giulianotti, R. and Klauser, F. (2010). Security governance and sport mega-events: Toward an interdisciplinary research agenda. *Journal of Sport & Social Issues,* 34(1), 49–61.

Goffman, E. (1957). Alienation from interaction. *Human Relations,* 10(1), 47–60.

Goold, B., Loader, I. and Thumala, A. (2010). Consuming security? Tools for a sociology of security consumption. *Theoretical Criminology,* 14(3), 3–30.

Graham, S. (2005). Software-sorted geographies. *Progress in Human Geography,* 29(5), 562–580.

Graham, S. (2012). Olympics 2012 security: Welcome to lockdown London. *City,* 16(4), 446–451.

Graham, S. and Thrift, N. (2007). Out of order: Understanding repair and maintenance. *Theory, Culture & Society,* 24(3), 1–25.

Gray, B. (1985). Conditions facilitating interorganizational collaboration. *Human Relations,* 38(10), 911–936.

Grewal, R., Comer, J. M. and Mehta, R. (2001). An investigation into the antecedents of organizational participation in business-to-business electronic markets. *Journal of Marketing,* 65(3), 17–33.

Grimshaw, D., Vincent, S. and Willmott, H. (2002). Going privately: Partnership and outsourcing in UK public services. *Public Administration,* 80(3), 475–502.

Grugulis, I. and Vincent, S. (2009). Whose skill is it anyway? 'Soft' skills and polarization. *Work, Employment and Society,* 23(4), 597–615.

Grusin, R. A. (2004). Premediation. *Criticism,* 46(1), 17–39.

Haggerty, K. D. and Ericson, R. V. (2000). The surveillant assemblage. *The British Journal of Sociology,* 51(4), 605–622.

Hair, J. F., Black, W. C., Babin, B. J. and Anderson, R. E. (2009). *Multivariate Data Analysis* (7th edition). Upper Saddle River: Prentice Hall.

Hall, S. (1993). Introduction: For Allon White: Metaphors of transformation. In: Barrell, J., Rose, J., Stallybrass, P. and White, J. (eds.), *Carnival, Hysteria, and Writing.* New York: Oxford University Press, 1–25.

Hanlon, G. (2008). Re-thinking corporate social responsibility and the role of the firm: On the denial of politics. In: Crane, A., Matten, D., McWilliams, A., Moon, J. and Siegel, D. S. (eds.), *The Oxford Handbook of Corporate Social Responsibility.* Oxford: Oxford University Press.

Hanlon, G. and Fleming, P. P. (2009). Updating the critical perspective on corporate social responsibility. *Sociology Compass*, 3(6), 937–948.

Hardt, M. and Negri, A. (2000). *Empire.* Cambridge: Harvard University Press.

Hardy, C., Phillips, N. and Clegg, S. (2001). Reflexivity in Organization and Management Theory: A Study of the Production of the Research 'Subject'. *Human Relations*, 54(5), 531–560.

Harvey, J. (2005). An evaluation of money laundering policies. *Journal of Money Laundering Control*, 8(4), 339–345.

Hebson, G., Grimshaw, D. and Marchington, M. (2003). PPPs and the changing public sector ethos: Case-study evidence from the health and local authority sectors. *Work, Employment and Society*, 17(3), 481–501.

Henfridsson, O. and Bygstad, B. (2013). The generative mechanisms of digital infrastructure evolution. *Management Information Systems Quarterly*, 37(3), 896–931.

Henneberg, S. C. (2005). An exploratory analysis of CRM implementation models. *Journal of Relationship Marketing*, 4(3/4), 85–104.

Henson, S. and Heasman, M. (1998). Food safety regulation and the firm: Understanding the compliance process. *Food Policy*, 23(1), 9–23.

Hepsø, V., Monteiro, E. and Rolland, K. H. (2009). Ecologies of e-infrastructures. *Journal of the Association for Information Systems*, 10(5), 430–446.

Herrmann, G. M. (2003). Negotiating culture: Conflict and consensus in US garage-sale bargaining. *Ethnology*, 42(3), 237–252.

Hirsch, E. (1967). *Validity in Interpretation.* New Haven: Yale University Press.

Hitchings, R. (2003). People, plants and performance: On actor network theory and the material pleasures of the private garden. *Social & Cultural Geography*, 4(1), 99–114.

Hochschild, A. (1983). *The Management Heart: The Commercialisation of Human Feeling.* Berkeley: University of California Press.

Homburg, C., Grozdanovic, M. and Klarmann, M. (2007). Responsiveness to customers and competitors: The role of affective and cognitive organizational systems. *Journal of Marketing*, 71(3), 18–38.

Home Affairs Committee (2009). *The eBorders Programme*, available from http://www.publications.parliament.uk/pa/cm200910/cmselect/cmhaff/170/17002.htm, accessed 15 Jan 2010.

Home Affairs Committee (2012). *The Work of the Border Force*, available from http://www.publications.parliament.uk/pa/cm201213/cmselect/cmhaff/523/523.pdf, accessed 17 July 2012.

Home Affairs Committee (2012). *Home Affairs Committee announces inquiry into e-Crime*, available from http://www.parliament.uk/business/committees/committees-a-z/commons-select/home-affairs-committee/news/120601-e-crime-call-for-ev/, accessed December 2012.

Home Office (2011). *Prevent Strategy*, https://www.gov.uk/government/uploads/system/uploads/attachment_data/file/97976/prevent-strategy-review.pdf, accessed September 2011.

Hood, B. M., Rothstein, H. and Baldwin, R. (2001). *The Governance of Risk*. Oxford: Oxford University Press.

Howcroft, D. and Richardson, H. (2012). The back office goes global: Exploring connections and contradictions in shared service centres. *Work, Employment and Society*, 26(1), 111–127.

Hutter, B. M. (2006). *The Role of Non-State Actors in Regulation*. The Centre for Analysis of Risk and Regulation, London School of Economics. http://www.lse.ac.uk/researchAndExpertise/units/CARR/pdf/DPs/Disspaper37.pdf, accessed June 2009.

Hutter, B. M. and Jones, C. J. (2007). From government to governance: External influences on business risk management. *Regulation & Governance*, 1(1), 27–45.

Huysmans, J. (2006). *The Politics of Insecurity: Fear, Migration and Asylum in the EU*. London: Routledge.

Huysmans, J. (2011). What's in an act? On security speech acts and little security nothings. *Security Dialogue*, 42(4–5), 371–383.

Huysmans, J. (2014). *Security Unbound: Enacting Democractic Limits*. London: Routledge.

Huysmans, J. and Buonfino, A. (2008). Politics of Exception and Unease: Immigration, Asylum and Terrorism in Parliamentary Debates in the UK. *Political Studies*, 56(4), 766–788.

IFAC (2004). *Anti-Money Laundering.* New York: International Federation of Accountants.

Jaakkolaa, E. and Hakanen, T. (2013). Value co-creation in solution networks. *Industrial Marketing Management*, 42(1), 47–58.

Jamal, T. B. and Getz, D. (1995). Collaboration theory and community tourism planning. *Annals of Tourism Research*, 22(1), 186–204.

Johnson, D. S., Clark, B. H. and Barczak, G. (2012). Customer relationship management processes: How faithful are business-to-business firms to customer profitability? *Industrial Marketing Management*, 41(7), 1094–1105.

Johnston, L. and Shearing, C. (2003). *Governing Security: Explorations in Policing and Justice*. London: Routledge.

Joint Money Laundering Steering Group (2006). *Prevention of Money Laundering/ Combatting Terrorist Financing.* http://www.jmlsg.org.uk/industry-guidance/ article/guidance, accessed February 2009.

Kagan, R. A. (2006). Environmental management style and corporate environmental performance. In: Coglianese, C. and Nash, J. (eds.), *Leveraging the Private Sector.* Washington: RFF Press, 31–50.

Kallinikos, J. (2005). The order of technology: Complexity and control in a connected world. *Information and Organization*, 15(3), 185–202.

Kang, G. and James, J. (2007). Revisiting the concept of a societal orientation: Conceptualization and delineation. *Journal of Business Ethics*, 73(3), 301–318.

Karakostas, B., Kardaras, D. and Papathanassiou, E. (2005). The state of CRM adoption by the financial services in the UK: an empirical investigation. *Information & Management*, 42(6), 853–863.

Keramati, A., Mehrabi, H. and Mojir, N. (2010). A process-oriented perspective on customer relationship management and organizational performance: An empirical investigation. *Industrial Marketing Management*, 39(7), 1170–1185.

Kingdon, J. (2004). AI Fights Money Laundering. *Ieee Intelligent Systems & Their Applications*, 19(3), 87–89.

Knights, D. and Morgan, G. (eds.) (1997). *Regulation and Deregulation in European Financial Services.* London: Macmillan.

Knights, D. and Murray, F. (1994). *Managers Divided: Organisation Politics and Information Technology Management.* Chichester: John Wiley & Sons, Inc.

Kollewe, J. (2010). Home office strips Raytheon Group of border control contract. *The Guardian*, 23 July.

Koskela, H. (2000). 'The gaze without eyes': video-surveillance and the changing nature of urban space. *Progress in Human Geography*, 24(2), 243–265.

Kotler, P. (2012). Marketing Needs a Conscience. *American Marketing Association Marketing News*, 46, 30.

Krasnikov, A. and Jayachandran, S. (2008). The Relative Impact of Marketing, Research-and-Development, and Operations Capabilities on Firm Performance. *Journal of Marketing*, 72(4), 1–11.

Kumar, N., Stern, L. W. and Anderson, J. C. (1993). Conducting interorganizational research using key informants. *The Academy of Management Journal*, 36(6), 1633–1651.

Lai, K., Wong, C. W. Y. and Cheng, T. C. E. (2010). Bundling digitized logistics activities and its performance implications. *Industrial Marketing Management*, 39(2), 273–286.

Laja, S. (2012). Long-awaited NHS information strategy 'could mean better data linking. *The Guardian*, 3 May. www.guardian.co.uk/government-computing-network/2012/May/03/nhs-information-strategy-data-linking, accessed 5 Sept 2012.

Latour, B. (2005). *Reassembling the Social: An Introduction to Actor-Network Theory*. Oxford: Oxford University Press.

Laurier, E. (2001). Why people say where they are during mobile phone calls. *Environment and Planning D*, 19(4), 485–504.

Laurier, E. and Philo, C. (2006). *Possible Geographies: a passing encounter in a café*, available from www.geos.ed.ac.uk/homes/elaurier/texts/Laurier_CV.pdf, accessed January 2014.

Lazzarato, M. (2006). Life and living in the societies of control. In: Fuglsang, M. and Sørensen, B. M. (eds.), *Deleuze and the Social*. Edinburgh: Edinburgh University Press, 171–190.

Lees, L. and Baxter, R. (2011). A 'building event' of fear: Thinking through the geography of architecture. *Social & Cultural Geography*, 12(2), 107–122.

Leidner, R. (1999). Emotional labor in service work. *The ANNALS of the American Academy of Political and Social Science*, 561(1), 81–95.

Levi, M. and Gilmore, B. (2002). Terrorist finance, money laundering and the rise and rise of mutual

evaluation: a new paradigm for crime control? *European Journal of Law Reform*, 4(2), 337–364.

Levi, M. and Wall, D. S. (2004). Technologies, Security, and Privacy in the Post-9/11 European Information Society. *Journal of Law and Society*, 31(2), 194–220.

Levi-Faur, D. (2005). The global spread of regulatory capitalism. *The ANNALS of the American Association of Political and Social Sciences*, 598, 12–32.

Levinas, E. (1986). Reality and its shadow. In: Levinas, E., *Collected Philosophical Papers*. Netherlands: Springer, 1–13.

Leyshon, A. and Thrift, N. (1999). Lists come alive: Electronic systems of knowledge and the rise of credit-scoring in retail banking. *Economy and Society*, 28(3), 434–466.

Liberty (2012). http://www.liberty-human-rights.org.uk/pdfs/policy12/liberty-submission-to-the-draft-communications-data-bill-committee-aug-2012-.pdf, accessed June 2013.

Lin, Y., Su, H. and Chien, S. (2006). A knowledge-enabled procedure for customer relationship management. *Industrial Marketing Management*, 35, 446–456.

Lindgreen, A., Hingley, M. K., Grant, D. B. and Morgan, R. E. (2012). Value in business and industrial marketing: Past, present, and future. *Industrial Marketing Management*, 41(1), 207–214.

Lindgreen, A., Palmer, R., Vanhamme, J. and Wouters, J. (2006). A relationship-management assessment tool: Questioning, identifying, and prioritizing critical aspects of customer relationships. *Industrial Marketing Management*, 35(1), 57–71.

Loader, I. and Walker, N. (2010). *Civilising Security*. Cambridge: Cambridge University Press.

Longfellow, T. (2006). Compliance tips regarding anti-money laundering. *Journal of Financial Planning*, Sep2006 Supplement, 15.

Luhmann, N. (1993). *Risk: A Sociological Theory*. New Brunswick: Transaction Publishers.

Lyon, D. (1994). *The Electronic Eye: The Rise of Surveillance Society*. Minneapolis: University of Minnesota Press.

Lyon, D. (2001). *Surveillance Society: Monitoring Everyday Life*. Buckingham: Open University Press.

Lyon, D. (2002). Everyday surveillance: Personal data and social classifications. *Information, Communication and Society*, 5(2), 242–257.

Lyon, D. (2009). *Identifying Citizens: ID Cards as Surveillance*. Cambridge: Polity Press.

Lyon, D., Haggerty, K. and Ball, K. (2012). Introducing Surveillance Studies. In: Ball, K., Haggerty, K. and Lyon, D. (eds.), *The Routledge Handbook of Surveillance Studies*. London: Routledge.

MacCormack, T. (2001). Let's get personal: Exploring the professional persona in health care. *The Qualitative Report 6*. http://www.nova.edu/ssss/QR/QR6-3/maccormack.html, accessed 10 September 2012.

MacLeod, V. and McLindin, B. (2011). Methodology for the evaluation of an international airport automated border control processing system. In: Jain, L. C.,

Aidman, E. V. and Abeynayake, C. (eds.), *Innovations in Defence Support Systems – 2*. Berlin, Heidelberg: Springer, 115–145.

Mantzana, V., Themistocleous, M., Irani, Z. and Morabito, V. (2006). Identifying healthcare actors involved in the adoption of information systems. *European Journal of Information Systems*, 16(1), 91–102.

Marx, G. T. and Muschert, G. W. (2007). Personal information, borders, and the new surveillance studies. *Annual Review of Law and Social Science*, 3(1), 375–395.

Marx, K. (1976). *Capital, Volume 1*. Harmondsworth: Penguin.

Massey, C., Alpass, F., Flett, R., Lewis, K., Morriss, S. and Sligo, F. (2006). Crossing fields: The case of a multi-disciplinary research team. *Qualitative Research*, 6(2), 131–147.

McCahill, M. and Finn, R. (2014). *Surveillance Capital and Resistance: Theorising the surveillance Subject*. London: Routledge.

McCue, C. (2006). *Data mining and predictive analysis – Intelligence gathering and crime analysis*. Oxford: Butterworth-Heinemann.

Meadows, M. and Dibb, S. (2012). Progress in customer relationship management adoption: a cross-sector study. *Journal of Strategic Marketing*, 20(4), 323–344.

Metro-Roland, D. (2010). Hip hop hermeneutics and multicultural education: A theory of cross-cultural understanding. *Educational Studies*, 46(6), 560–578.

Michaels, J. D. (2010). Deputizing Homeland Security. *Texas Law Review*, 88(7), 1435.

Midgley, G. (2011). Theoretical pluralism in systemic action research. *Systemic Practice and Action Research*, 24(1), 1–15.

Miles, M. B. and Huberman, A. M. (1994). *Qualitative Data Analysis*. Thousand Oaks: Sage.

Millward, D. (2009). Border chaos fear. *The Daily Telegraph*, 18 April.

Mitchell, R., Agle, B. and Wood, D. (1997). Toward a theory of stakeholder identification and salience: Defining the principle of who and what really counts. *Academy of Management Review*, 22(4), 853–886.

Monahan, T. (2011). Surveillance as cultural practice. *The Sociological Quarterly*, 52(4), 495–508.

Monahan, T. and Palmer, N. A. (2009). The emerging politics of DHS fusion centers. *Security Dialogue*, 40(6), 617–636.

Morgan, N. A., Vorhies, D. W. and Mason, C. H. (2009). Market orientation, marketing capabilities, and firm performance. *Strategic Management Journal*, 30(8), 909–920.

Morgan, R. M. and Hunt, S. D. (1994). The Commitment-Trust Theory of Relationship Marketing. *Journal of Marketing*, 58(3), 20–38.

Nalla, M. K. and Hwang, E.-G. (2006). Relations between police and private security officers in South Korea. *Policing: An International Journal of Police Strategies and Management*, 29 (3), 482–497.

Newbury, J. (2011). A place for theoretical inconsistency. *International Journal of Qualitative Methods*, 10(4), 335–347.

Neyland, D. (2007). *Organizational Ethnography*. London: Sage.

Nordin, F., Kindström, D., Kowalkowski, C. and Rehme, J. (2011). The risks of providing services: Differential risk effects of the service-development strategies of customisation, bundling, and range. *Journal of Service Management*, 22(3), 390–408.

Norris, C. and Armstrong, G. (1999). *The Maximum Surveillance Society: The Rise of CCTV*. Oxford and New York: Berg Publishers.

Norris, C. and McCahill, M. (2006). CCTV: Beyond penal modernism? *British Journal of Criminology*, 46(1), 97–118.

Norusis, M. J. (2002). *SPSS guide to data analysis*. Upper Saddle River: Prentice Hall.

Orlikowski, W. J. (1992). The Duality of Technology: Rethinking the Concept of Technology in Organizations. *Organization Science*, 3(3), 398–427.

OTA (1995). *Technologies for the Control of Money Laundering*. Washington DC: Congress Office of Technology Assessment, 157.

Pan, G. (2005). Information systems project abandonment: A stakeholder analysis. *International Journal of Information Management*, 25(2), 173–184.

Parker, C. and Nielsen, V. (2009). The challenge of empirical research on business compliance in regulatory capitalism. *Annual Review of Law and Social Science*, 5, 45–70.

PATS project (2011). *Privacy Awareness through Security Organisation Branding*. http://www.pats-project.eu/, accessed November 2011.

Payne, A. and Frow, P. (2005). A strategic framework for customer relationship management. *Journal of Marketing*, 69(4), 167–176.

Phillips, R. (2003). Stakeholder legitimacy. *Business Ethics Quarterly*, 13(1), 25–41.

Polites, G. and Karahanna, E. (2013). The embeddedness of information systems habits in organizational and individual level routines: Development and disruption. *MIS Quarterly*, 37(1), 221–246.

Porter, M. E. (2011). *Competitive advantage of nations: Creating and sustaining superior performance*. New York: Free Press.

Porter, M. E. and Kramer, M. R. (2011). Creating Shared Value. *Harvard Business Review*, Jan-Feb, 62-77.

Pouloudi, A. and Whitley, E. A. (1997). Stakeholder identification in inter-organizational systems: Gaining insights for drug use management systems. *European Journal of Information Systems*, 6(1), 1–14.

Quack, S. and Hildebrandt, S. (1997). Bank finance for small and medium-sized enterprises in Germany and France. In: Morgan, G. and Knights, D. (eds.), *Regulation and Deregulation in European Financial Services*. London: Macmillan.

Ravenscroft, N. and Gilchrist, P. (2009). Spaces of transgression: Governance, discipline and reworking the carnivalesque. *Leisure Studies*, 28(1), 35–49.

Reals, K. (2008). Airlines prepare for new UK border control initiative. *Flightglobal*, 24 (November), available from http://www.flightglobal.com/news/articles/airlines-prepare-for-new-uk-border-control-initiative-319332/, accessed 4 March 2009.

Reed, D. (2002). Employing normative stakeholder theory in developing countries: A critical theory perspective. *Business and Society*, 41(2), 166–207.

Ribes, D. and Finholt, T. (2009). The long now of technology infrastructure: Articulating tensions in development. *Journal of Applied Information Systems*, 10(5), 375–398.

Ribes, D., Jackson, S., Geiger, S., Burton, M. and Finholt, T. (2013). Artefacts that organize: Delegation in the distributed organization. *Information and Organization*, 23(1), 1–14.

Richard, J. E., Thirkell, P. C. and Huff, S. L. (2007). An Examination of Customer Relationship Management (CRM) Technology Adoption and its Impact on Business-to-Business Customer Relationships. *Total Quality Management & Business Excellence*, 18(8), 927–945.

Richards, N. M. (2013). The dangers of surveillance. *Harvard Law Review*, 126(7), 1934–1965.

Rigakos, G. (2008). *Nightclub: Bouncers, Risk and the Spectacle of Consumption*. Montreal and Kingston: McGill-Queen's University Press.

Rivard, S. and Lapointe, L. (2012). Information technology implementers' responses to user resistance: Nature and effects. *MIS Quarterly*, 36(3), 897–927.

Robinson, J. P., Shaver, P. R. and Wrightsman, L. S. (1991). Criteria for scale selection and evaluation. In: Robinson, J. P., Shaver, P. R. and Wrightsman, L. S. (eds.), *Measures of Personality and Social Psychological Attitudes*. San Diego: Academic Press.

Rollins, M., Bellenge, D. N. and Johnston, W. J. (2012). Customer information utilization in business-to-business markets: Muddling through process? *Journal of Business Research*, 65(6), 758–764.

Roper, I., James, P. and Higgins, P. (2005). Workplace partnership and public service provision: The case of the 'best value' performance regime in British local government'. *Work, Employment and Society*, 19(3), 639–649.

Ryals, L. and Payne, A. (2001). Customer relationship management in financial services: Towards information-enabled relationship marketing. *Journal of Strategic Marketing*, 9(1), 3–27.

Ryder, N. (2008). The financial services authority and money laundering. *The Cambridge Law Journal*, 67(3), 635–653.

Sahay, S., Aanestad, M. and Monteiro, E. (2009). Configurable politics and asymmetric integration: Health e-Infrastructures in India. *Journal of the Association for Information Systems*, 10(5), 399–141.

Salomon, D. (1999). Deregulation and embeddedness: The case of the French banking system. In: Morgan, G. and Engwall, L. (eds.), *Regulation and Organization: International Perspectives*. London: Routledge, 69–81.

Salter, M. B. (2004). Passports, mobility, and security: How smart can the border be? *International Studies Perspectives*, 5(1), 71–91.

Samatas, M. (2005). Studying surveillance in Greece: Methodological and other problems related to an authoritarian surveillance culture. *Surveillance & Society*, 3(2/3), 181–197.

Sandner, D. (2004). *Fantastic Literature: A Critical Reader*. Westport: Praeger Publishers.

Saunders, M., Lewis, P. and Thornhill, A. (2007). *Research Methods for Business Students*. London: Prentice Hall.

Schneier, B. (2014). Don't listen to Google and Facebook: The Public-Private Surveillance Partnership is Still Going Strong. *The Atlantic*, 25 March. http://www.theatlantic.com/technology/archive/2014/03/don-t-listen-to-google-and-facebook-the-public-private-surveillance-partnership-is-still-going-strong/284612/, accessed March 20.

Sethi, R. (2000). New Product Quality and Product Development Teams. *Journal of Marketing*, 64(2), 1–14.

Sennett, R. (ed.) (2006). *The Culture of the New Capitalism*. Boston: Yale University Press.

Sharman, J. C. and Chaikin, D. (2009). *Corruption and Money Laundering: A Symbiotic Relationship*. New York: Palgrave.

Shaw, G., Bailey, A. and Williams, A. (2011). Aspects of service-dominant logic and its implications for tourism management: examples from the hotel industry. *Tourism Management*, 32(2), 207–214.

Sheth, J. N., Sisodia, R. S. and Sharma, A. (2000). The antecedents and consequences of customer-centric marketing. *Journal of the Academy of Marketing Science,* 28(1), 55–66.

Shum, P., Bove, L. and Auh, S. (2008). Employees' affective commitment to change: The key to successful CRM implementation. *European Journal of Marketing,* 42(11–12), 1346–1371.

Sica, V. (2000). Cleaning the laundry: States and the monitoring of the financial system. *Millennium-Journal of International Studies,* 29(1), 47–72.

Skinns, L. (2008). A prominent participant? The role of the state in police partnerships: REVIEW ESSAY. *Policing & Society,* 18(3), 311–321.

Smart, C. and Vertinsky, I. (1984). Strategy and the environment: a study of corporate responses to crises. *Strategic Management Journal,* 5(3), 199–213.

Smith, A. (2007). Emerging in between: The multi-level governance of renewable energy in the English regions. *Energy Policy,* 35(12), 6266–6280.

Smith, A. (2012). 'Monday will never be the same again': The transformation of employment and work in a public-private partnership. *Work, Employment & Society,* 26(1), 95–110.

Smith, A. M. (2012). Internal social marketing: lessons from the field of services marketing. In: Hastings, G., Angus, K. and Bryant, C. A. (eds.) (2012), *The Sage Handbook of Social Marketing.* Thousand Oaks: Sage, 298–316.

Snyder Bennear, L. (2006). Evaluating management based regulation: A valuable too in the regulatory toolbox? In: Coglianese, C. and Nash, J. (eds.), *Leveraging the Private Sector.* Washington: RFF Press, 51–86.

Stallybrass, P. and White, W. (1986). *The Politics and Poetics of Transgression.* New York: Cornell University Press.

Star, S. L. (1991). Power, technology and the phenomenology of conventions. In: Law, J. (ed.), *A Sociology Of Monsters. Essays on Power, Technology and Domination. Sociological Review Monograph,* 38. London: Routledge, 26–56.

Star, S. L. (1999). The Ethnography of Infrastructure. *American Behavioral Scientist,* 43(3), 377–391.

Star, S. L. and Griesemer, J. R. (1989). Institutional ecology, 'translations' and boundary objects: Amateurs and professionals in Berkeley's Museum of Vertebrate Zoology, 1907–39. *Social Studies of Science,* 19(3), 387–420.

Star, S. L. and Ruhleder, K. (1996). Steps toward an ecology of infrastructure: Design and access for large information systems. *Information Systems Research,* 7(1), 111–134.

Stein, A. D., Smith, M. F. and Lancioni, R. A. (2013). The development and diffusion of customer relationship management (CRM) intelligence in business-to-business environments. *Industrial Marketing Management*, 42(6), 855–861.

Stürmer, S. and Simon, B. (2009). Pathways to collective protest: Calculation, identification, or emotion? A critical analysis of the role of group-based anger in social movement participation. *Journal of Social Issues*, 65(4), 681–705.

Surveillance Studies Network (2006). *A Report on the Surveillance Society.* Wilmslow: Information Commissioner's Office.

Surveillance Studies Network (2010). An Update to a report on the surveillance society UK Information Commissioner. Wilmslow: Information Commissioner's Office.

Tashakkori, A. and Teddlie, C. (2003). *Handbook of Mixed Methods in Social and Behavioural Research.* Thousand Oaks: Sage.

Taylor, C. (2007). *Modern Social Imaginaries.* London: Duke University Press.

Taylor, P. and Cooper, C. (2008). 'It was absolute hell': Inside the private prison. *Capital and Class*, 96, 3–30.

Theodosiou, M., Kehagias, J. and Katsikea, E. (2012). Strategic Orientations, marketing capabilities and firm performance: An empirical investigation in the context of frontline managers in service organizations. *Industrial Marketing Management*, 41(7), 1058–1070.

Thompson, P., Warhurst, C. and Callaghan, G. (2001). Ignorant theory and knowledgeable workers: Interrogating the connections between knowledge, skills and services. *Journal of Management Studies*, 38(7), 923–942.

Thumala, A., Goold, B. and Loader, I. (2011). A tainted trade? Moral ambivalence and legitimation work in the private security industry. *The British Journal of Sociology*, 62(2), 283–303.

Tosun, C. (2000). Limits to community participation in the tourism development process in developing countries. *Tourism Management*, 21(6), 613–633.

Trevino, L. and Weaver, G. (1999). The stakeholder research tradition: Converging theorists – not convergent theory. *Academy of Management Review*, 24(2), 222–227.

Troy, L. C., Hirunyawipada, T. and Paswan, A. K. (2008). Cross-functional integration and new product success: An empirical investigation of the findings. *Journal of Marketing*, 72(6), 132–146.

Turow, J. (2006). Cracking the consumer code: Advertising, anxiety and surveillance in the Digital Age. In: Haggerty, K. and Ericson, R. (eds.), *The New Politics of Surveillance and Visibility.* London: Routledge, 279–307.

Turow, J. (2012). *The Daily You: How the New Advertising Industry is Defining Your Identity and Your Worth.* Boston: Yale University Press.

Tyler, T. R. and Blader, S. L. (2005). Can businesses effectively regulate employee conduct? The antecedents of rule following in work settings. *Academy of Management Journal*, 48(6), 1143–1158.

Ure, J., Procter, R., Lin, Y-W., Hartswood, M., Anderson, S., Lloyd, S., Wardlaw, J., Gonzalez-Velez, H. and Ho, K. (2009). The Development of Data Infrastructures for eHealth: A Socio-Technical Perspective. *Journal of the Association for Information Systems*, 10(5), 415–429.

Vakalis, I., Camenisch, J., Leenes, R. and Sommer, D. (2011). Airport security controls: Digital privacy. In: Camenisch, J., Leenes, R. and Sommer, D. (eds.), *Digital Privacy: Privacy and Identity Management in Europe.* Berlin and Heidelberg: Springer, 721–734.

Van Calster, P. J. (2011). Privatising criminal justice? Shopping in the Netherlands. *The Journal of Criminal Law*, 75(3), 204–224.

Vandermerwe, S. (2004). Achieving Deep Customer Focus. *MIT Sloan Management Review*, 45(3), 26–34.

Van de Ven, A. H. and Poole, M. S. (2005). Alternative approaches for studying organizational change. *Organization Studies*, 26(9), 1377–1404.

Vernon, J., Essex, S., Pinder, D. and Curry, K. (2005). Collaborative policymaking. Local sustainable projects. *Annals of Tourism Research*, 32(2), 325–345.

Vikkelsø, S. (2007). In between curing and counting: Performative effects of experiments with healthcare information infrastructure. *Financial Accountability & Management*, 23(3), 269–288.

Vlcek, W. (2007). Surveillance to combat terrorist financing in Europe: Whose liberty, whose security? *European Studies*, 16(1), 99–119.

Wagenaar, P. and Boersma, K. (2012). Zooming in on 'heterotopia': CCTV-operator practices at Schiphol Airport. *Information Polity*, 17(1), 7–20.

Wakefield, A. (2003). *Selling Security: The Private Policing of Public Space.* Cullompton: Willan.

Ward, J. M., Hemingway, C. and Daniel, E. M. (2005). A framework for addressing the organisational issues of enterprise systems implementation. *Journal of Strategic Information Systems*, 14(2), 97–119.

Weaver, A. (2005). Interactive service work and performative metaphors: The case of the cruise industry. *Tourist Studies*, 5(1), 5–27.

Webb, L. (2004). A survey of money laundering reporting officers and their attitudes towards money laundering regulations. *Journal of Money Laundering Control*, 7(4), 367–375.

Westley, F. and Vredenburg, H. (1991). Strategic bridging: The collaboration between environmentalists and business in the marketing of green products. *Journal of Applied Behavioural Science*, 27(1), 65–90.

White, A. (1993). *Carnival, Hysteria, and Writing.* New York: Oxford University Press.

White, A. (2011). The new political economy of private security. *Theoretical Criminology*, 16(1), 85–101.

Whitehead, T. (2009). £1.2billion eBorders plan left in disarray. *The Daily Telegraph*, 18 December, 12.

Whitley, E. A. (2008). Perceptions of government technology, surveillance and privacy: The UK identity cards scheme. In: Goold, B. J. and Neyland, D. (eds.), *New Directions in Privacy and Surveillance.* London: Routledge, 133–156.

Woolas, P. (2009). Hi-tech controls will trap bogus students: Anyone who wants to see Britain's borders protected from potential terrorists should welcome the new immigration plans. *The Times*, 17 April, 19.

Yeandle, M., Mainelli, M., Berendt, A. and Healy, B. (2005). Anti money laundering requirements: Costs, benefits and perceptions. *City Research Series.* London: Z/Yen.

Zablah, A. R., Bellenger, D. N. and Johnston, W. J. (2004). An evaluation of divergent perspectives on customer relationship management: Towards a common understanding of an emerging phenomenon. *Industrial Marketing Management*, 33(6), 475–489.

Zdanowicz, J. S. (2004). Money laundering and terrorist financing. *Communications of the ACM*, 47(5), 53–55.

Appendices

Appendix A: Correlations between selected questions (within each main section of the survey)

Tables A1–A5 show significant correlations within each main section of the survey.

* indicates that the correlation is significant at the 0.05 level (2-tailed);

** indicates that the correlation is significant at the 0.01 level (2-tailed)

		Q1	Q2	Q3	Q4
Q1: There is no stated desire at all within the organisation for RM	Pearson Correlation	1	.584**	.690**	.496**
	Sig. (2-tailed)		.000	.000	.000
Q2: CRM does not have a strong champion at the top of the organisation	Pearson Correlation		1	.840**	.644**
	Sig. (2-tailed)			.000	.000
Q3: Senior management is not at all proactive in supporting CRM projects	Pearson Correlation				.668**
	Sig. (2-tailed)				.000
Q4: The organisational culture is not well suited to supporting CRM	Pearson Correlation				1
	Sig. (2-tailed)				

Table A1. Correlations between 'Pre-requisites for CRM' questions.

		Q1	Q2	Q3	Q4	Q5
Q1: The emphasis is on using information to record transactions rather than as a strategic tool	Pearson Correlation	1	.238*	.200	.400**	.186
	Sig. (2-tailed)		.043	.088	.000	.115
Q2: Front-line staff have access to only very basic customer data when handling customer enquiries	Pearson Correlation		1	.409**	.322**	.235*
	Sig. (2-tailed)			.000	.006	.044
Q3: Computer system design and implementation are driven by internal accounting needs rather than external customer needs	Pearson Correlation			1	.172	.232*
	Sig. (2-tailed)				.142	.047
Q4: Systems do not have access to attitudinal/buying behavior data required to identify 'life events'	Pearson Correlation				1	.261*
	Sig. (2-tailed)					.026
Q5: Those handling customer direct marketing never co-ordinate their activities with front-line staff	Pearson Correlation					1
	Sig. (2-tailed)					

Table A2. Correlations between 'Technology' questions around CRM implementation.

		Q1	Q2	Q3
Q1: Emphasis is on the value to be achieved from customers today (perhaps through the sale of an additional product) rather than on customers' life-time value	Pearson Correlation	1	.392**	.464**
	Sig. (2-tailed)		.001	.000
Q2: During customer contact the emphasis is always on conducting transactions rather than on updating customer information systems	Pearson Correlation		1	.314**
	Sig. (2-tailed)			.008
Q3: The company is very poor at anticipating and reacting to customer needs (events-based marketing)	Pearson Correlation			1
	Sig. (2-tailed)			

Table A3. Correlations between 'Customer' questions around CRM implementation.

		Q1	Q2	Q3	Q4
Q1: Emphasis is on transaction-driven marketing rather than customer-driven/life event-led marketing	Pearson Correlation	1	.443**	.508**	.571**
	Sig. (2-tailed)		.000	.000	.000
Q2: The company always focuses on customer groups rather than the individual	Pearson Correlation		1	.340**	.395**
	Sig. (2-tailed)			.003	.000
Q3: The company focuses on increasing sales volumes rather than relationship building as the route to competitive advantage	Pearson Correlation			1	.503**
	Sig. (2-tailed)				.000
Q4: CRM implementation does not permeate all parts of the organisation	Pearson Correlation				1
	Sig. (2-tailed)				

Table A4. Correlations between 'Company' questions around CRM implementation.

		Q1	Q2
Q1: Senior management never sets objectives which reflect the company stance on CRM	Pearson Correlation	1	.466**
	Sig. (2-tailed)		.000
Q2: Staff never use day-to-day contacts with customers as a market research opportunity	Pearson Correlation		1
	Sig. (2-tailed)		

Table A5. Correlations between 'Staff' questions around CRM implementation.

Appendix B: Correlations between selected questions (across sections of the survey)

Tables B1–B5 show correlations between key questions in different sections of the survey.

* indicates that the correlation is significant at the 0.05 level (2-tailed);

** indicates that the correlation is significant at the 0.01 level (2-tailed)

		Q1	Q2	Q3	Q4	Q5	Q6	Q7	Q8	Q9	Q10	Q11	Q12
Q1. There is no stated desire at all within the organisation for RM	Pearson Correlation	1	.584**	.690**	.496**	-.276*	-.369**	-.192	-.224	-.178	-.191	-.210	-.167
	Sig. (2-tailed)		.000	.000	.000	.025	.002	.126	.066	.156	.127	.090	.183
Q2. CRM does not have a strong champion at the top of the organisation	Pearson Correlation		1	.840**	.644**	-.295*	-.268*	-.340**	-.209	-.191	-.140	-.130	-.075
	Sig. (2-tailed)			.000	.000	.016	.027	.006	.087	.128	.266	.296	.551
Q3. Senior management is not at all proactive in supporting CRM projects	Pearson Correlation			1	.668**	-.253*	-.342**	-.286*	-.239*	-.263*	-.293*	-.320**	-.113
	Sig. (2-tailed)				.000	.040	.004	.021	.050	.034	.018	.009	.372
Q4. The organisational culture is not well suited to supporting CRM	Pearson Correlation				1	-.197	-.175	-.294*	-.160	-.249*	-.163	-.331**	-.251*
	Sig. (2-tailed)					.113	.152	.017	.189	.045	.192	.006	.044
Q5. AML has changed the job role for those handling customer data	Pearson Correlation					1	.484**	.635**	.428**	.441**	.175	.227	.384**
	Sig. (2-tailed)						.000	.000	.000	.000	.163	.069	.002
Q6. Staff training has changed as a result of AML	Pearson Correlation						1	.273*	.519**	.376**	.386**	.390**	.504**
	Sig. (2-tailed)							.028	.000	.002	.001	.001	.000

Q7. AML has changed the job role of our front-line staff	Pearson Correlation							1	.218	.249*	.245	.313*	.155
	Sig. (2-tailed)								.081	.047	.051	.012	.220
Q8. Front-line staff have clear responsibility for reporting on unusual data that they observe	Pearson Correlation								1	.520**	.316**	.315**	.322**
	Sig. (2-tailed)									.000	.010	.009	.009
Q9. Our organisation has designed systems with AML in mind	Pearson Correlation									1	.332**	.434**	.198
	Sig. (2-tailed)										.007	.000	.113
Q10. AML projects are supported very strongly	Pearson Correlation										1	.838**	.204
	Sig. (2-tailed)											.000	.103
Q11. AML permeates all parts of the organisation	Pearson Correlation											1	.293*
	Sig. (2-tailed)												.018
Q12. The ways in which we analyse our data have changed as a result of AML	Pearson Correlation												1
	Sig. (2-tailed)												

Table B1. Correlations between AML and CRM pre-requisites.

		Q1	Q2	Q3	Q4	Q5	Q6	Q7	Q8	Q9
Q1. The emphasis is on using information to record transactions rather than as a strategic tool	Pearson Correlation	1	.238*	.400**	.186	.026	.043	-.180	-.196	-.297*
	Sig. (2-tailed)		.043	.000	.115	.835	.736	.151	.114	.017
Q2. Front-line staff have access to only very basic customer data when handling customer enquiries	Pearson Correlation		1	.322**	.235*	-.294*	-.291*	-.290*	-.380**	-.130
	Sig. (2-tailed)			.006	.044	.017	.019	.019	.002	.303
Q3. Systems do not have access to attitudinal/buying behaviour data required to identify 'life events'	Pearson Correlation			1	.261*	-.069	-.157	-.163	-.170	-.333**
	Sig. (2-tailed)				.026	.583	.216	.195	.173	.007
Q4. Those handling customer direct marketing never coordinate their activities with front-line staff	Pearson Correlation				1	-.138	-.180	-.109	-.253*	-.190
	Sig. (2-tailed)					.270	.151	.386	.040	.130
Q5. The ways in which we capture data have changed as a result of AML	Pearson Correlation					1	.789**	.168	.302*	.281*
	Sig. (2-tailed)						.000	.180	.014	.023
Q6. The ways in which we analyse our data have changed as a result of AML	Pearson Correlation						1	.204	.293*	.198
	Sig. (2-tailed)							.103	.018	.113

Q7. AML projects are supported very strongly	Pearson Correlation							1	.838**	.332**
	Sig. (2-tailed)								.000	.007
Q8. AML permeates all parts of the organisation	Pearson Correlation								1	.434**
	Sig. (2-tailed)									.000
Q9. Our organisation has designed systems with AML in mind	Pearson Correlation									1
	Sig. (2-tailed)									

Table B2. Correlations between AML and CRM implementation: The technology.

		Q1	**Q2**	**Q3**	**Q4**	**Q5**	**Q6**	**Q7**
Q1. During customer contact the emphasis is always on conducting transactions rather than on updating customer information systems	Pearson Correlation	1	.314**	.327**	.073	.319*	.252*	.277*
	Sig. (2-tailed)		.008	.009	.571	.011	.042	.029
Q2. The company is very poor at anticipating and reacting to customer needs (events-based marketing)	Pearson Correlation		1	.031	-.286*	.059	-.101	.216
	Sig. (2-tailed)			.806	.022	.639	.414	.087

Q3. The ways in which we capture data have changed as a result of AML	Pearson Correlation			1	.281*	.432**	.494**	.523**
	Sig. (2-tailed)				.023	.000	.000	.000
Q4. Our organisation has designed systems with AML in mind	Pearson Correlation				1	.441**	.376**	.137
	Sig. (2-tailed)					.000	.002	.276
Q5. AML has changed the job role for those handling customer data	Pearson Correlation					1	.484**	.460**
	Sig. (2-tailed)						.000	.000
Q6. Staff training has changed as a result of AML	Pearson Correlation						1	.527**
	Sig. (2-tailed)							.000
Q7. Our approach to dealing with customers has changed as a result of AML	Pearson Correlation							1
	Sig. (2-tailed)							

Table B3. Correlations between AML and CRM implementation: The customer.

		Q1	Q2	Q3	Q4	Q5	Q6
Q1. Emphasis is on transaction-driven marketing rather than customer-driven/life event-led marketing	Pearson Correlation	1	.443**	.508**	-.284*	.085	-.222
	Sig. (2-tailed)		.000	.000	.024	.509	.075
Q2. The company always focuses on customer groups rather than the individual	Pearson Correlation		1	.340**	-.032	.261*	.134
	Sig. (2-tailed)			.003	.803	.036	.281

Q3. The company focuses on increasing sales volumes rather than relationship building as the route to competitive advantage	Pearson Correlation			1	-.086	.025	-.330**
	Sig. (2-tailed)				.505	.848	.007
Q4. AML has changed the job role of our front-line staff	Pearson Correlation				1	.232	.313*
	Sig. (2-tailed)					.065	.012
Q5. Our approach to dealing with customers has changed as a result of AML	Pearson Correlation					1	.211
	Sig. (2-tailed)						.092
Q6. AML permeates all parts of the organisation	Pearson Correlation						1
	Sig. (2-tailed)						

Table B4. Correlations between AML and CRM implementation: The company.

		Q1	Q2	Q3	Q4	Q5
Q1: Senior management never sets objectives which reflect the company stance on CRM	Pearson Correlation	1	.466**	-.267*	-.404**	-.262*
	Sig. (2-tailed)		.000	.027	.001	.035
Q2: Staff never use day-to-day contacts with customers as a market research opportunity	Pearson Correlation		1	-.036	-.298*	-.044
	Sig. (2-tailed)			.773	.018	.734
Q3: Staff training has changed as a result of AML	Pearson Correlation			1	.323**	.335**
	Sig. (2-tailed)				.009	.006
Q4: The data accessible to front-line staff has changed as a result of AML	Pearson Correlation				1	.435**
	Sig. (2-tailed)					.000
Q5: The frequency of our customer contact has changed as a result of AML	Pearson Correlation					1
	Sig. (2-tailed)					

Table B5. Correlations between AML and CRM implementation: The staff.

Appendix C: AML competence and organisational pre-requisites for CRM

First we note the strong correlations between CRM pre-requisites on the one hand, and changing job roles due to AML on the other. This applies to changing job roles for both front-line staff and those handling customer data. Changing AML job roles for those handling customer data (Q5) is strongly associated with a stated desire for relationship management within the organisation (Q1; $r=0.276$, $p = 0.025$), as well as with having a strong CRM champion at the top of the organisation (Q2; $r=0.295$, $p = 0.016$) and with senior management being proactive in supporting CRM projects (Q3; $r=0.253$, $p = 0.04$).

It is not only the job role of *those handling customer data* that is related to CRM pre-requisites. Changing AML job roles *for front-line staff* (Q7) is also strongly associated with having a strong CRM champion at the top of the organisation, (Q2; $r=0.340$, $p = 0.006$) and with senior management being proactive in supporting CRM projects (Q3; $r=0.286$, $p = 0.021$). Q7 is also strongly correlated with an organisational culture that is well suited to supporting CRM (Q4; $r=0.294$, $p=0.017$).

In what ways is AML changing the role of staff, and how are these aspects associated with pre-requisites for CRM? We find that changes in the ways in which the data is being analysed due to AML (Q12) is associated with a particular pre-requisite for CRM – that the organisational culture is well suited to supporting CRM (Q4; $r=0.251$, $p=0.044$). Moreover, changing responsibilities for staff – in particular, the view that front-line staff have clear responsibility for reporting on unusual data that they observe (Q8) – is also associated with a particular pre-requisite for CRM – the notion that senior management are proactive in supporting CRM projects (Q3; $r=0.239$, $p = 0.05$).

It appears that changes in job role are reflected in changes in staff training and in systems for AML. Changes in staff training as a result of AML (Q6) are associated with a number of pre-requisites for CRM, namely a stated desire for relationship management within the organisation (Q1; $r=0.369$, $p = 0.002$), the presence of a strong CRM champion at the top of the organisation (Q2; $r=0.268$, $p = 0.027$) and the proactive support of senior management for CRM projects (Q3; $r=0.342$, $p = 0.004$). Furthermore, changes in systems – such as agreement with the proposition that the organisation has designed systems with AML in mind (Q9) – is associated with a number of pre-requisites for CRM, in particular the proactive support of senior management for CRM projects (Q3; r=0.263, p = 0.034), and the existence of an organisational culture that is well suited to supporting CRM (Q4; r=0.249, p=0.045).

Finally we note that there is a strong association between management support for CRM projects, and management support for AML projects – and that if the goal is for AML to permeate all parts of the organisation, both the support that management show for CRM and a culture that is well suited to CRM can help in the achievement of this goal. The proposition that AML projects are supported very strongly (Q10) is correlated with a particular pre-requisite for CRM – the proactive role of senior management in supporting CRM projects (Q3; r=0.293, p = 0.018). The notion that AML permeates all parts of the organisation (Q11) is associated with a number of pre-requisites for CRM – in particular, that senior management are proactive in supporting CRM projects (Q3; r=0.32, p = 0.009) and that the organisational culture is well suited to supporting CRM (Q4; r=0.331, p=0.006).

Appendix D: AML competence and the strategic deployment of technology for CRM purposes

First we note strong correlations between a particular technology-related issue – the access of staff to customer data when handling enquiries – in the implementation of CRM, and the ways in which AML has changed both data capture and data analysis. The propositions of change in both data capture (Q5) and data analysis (Q6) due to AML are strongly correlated with the statement that front-line staff have access to a range of customer data (rather than only very basic customer data) when handling customer enquiries (Q2) (for Q5, r = 0.294, p = 0.017; and for Q6, r = 0.291, p = 0.019). This suggests that two key areas of progress with AML – changes in data capture and data analysis – are supported by improved access to customer data as required by CRM projects.

Changes in data capture and analysis are likely to be reflected in changes to relevant systems. The idea that the organisation has designed systems with AML in mind (Q9) is associated with two items around CRM Implementation: The Technology – first, with the statement that the organisation emphasises using information as a strategic tool rather than simply to record transactions (Q1; $r = 0.297$, $p = 0.0170$) and second, with the statement that the organisation's systems have access to attitudinal/buying behaviour (Q3: $r = 0.333$, $p = 0.007$). This indicates that progress with CRM, such as the recognition of data as a strategic tool and associated improvements in the customer data that is available to staff, may support the provision of systems that are mindful of the needs of AML activity.

The idea that AML permeates all parts of the organisation (Q8) is associated with two questions around CRM Implementation: The Technology – first, the statement that front-line staff have access to a range of customer data when handling customer enquiries (Q2; $r = 0.38$, $p = 0.002$) and second, with the statement that staff handling customer direct marketing do tend to co-ordinate their activities with front-line staff (Q4; $r = 0.253$, $p = 0.04$). Finally, the idea that AML projects are supported very strongly (Q7) is associated with a particular question around CRM Implementation: The Technology – the proposition that front-line staff have access to a range of customer data when handling customer enquiries (Q2; $r = 0.290$, $p = 0.019$). Again, we see that successful AML may be supported by competence in CRM such as good practice in data access and data sharing.

Appendix E: AML competence and customer knowledge capabilities; organisational aspects of CRM and staff management and behaviours

Statistically significant correlations (as reported in Table B3, Appendix B) were found between two items representing customer-facing issues in the implementation of CRM (labelled Q1 and Q2) and five items representing key AML issues (labelled Q3 to Q7).

The notion that the organisation has designed systems with AML in mind (Q4) is associated with a particular question around CRM Implementation: The Customer – the company is good at anticipating and reacting to customer needs (events-based marketing) (Q2; $r = 0.286$, $p = 0.022$). There are also statistically significant correlations between the statement that during customer contact, the organisation tends to emphasise updating customer information systems rather than simply recording transactions (Q1), and four items representing key aspects of AML, namely the idea that

AML has changed the job role for those handling customer data (Q5; r = 0.319, p = 0.011); the notion that the ways in which data are captured have changed as a result of AML (Q3; r = 0.327, p = 0.009); the idea that staff training has changed as a result of AML (Q6; r = 0.252, p = 0.042); and the statement that the organisation's approach to dealing with customers has changed as a result of AML (Q7; r = 0.277, p = 0.029).

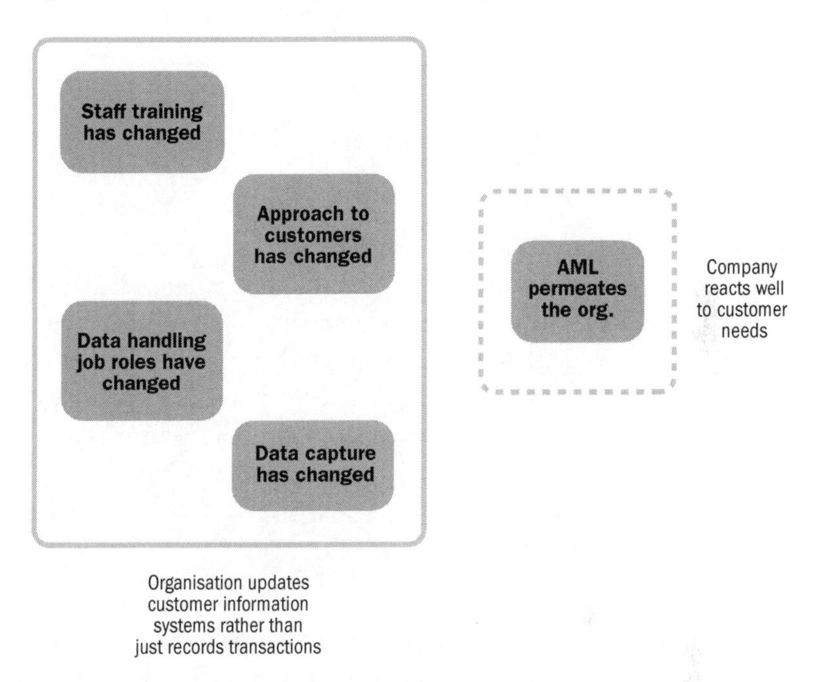

Figure 6.3. Customer knowledge capabilities correlations

Exploring the relationship between *AML competence* and the *organisation's strategic focus on customers*

Statistically significant correlations (as reported in Table B4, Appendix B) were found between three items representing key company issues in the implementation of CRM (labelled Q1 to Q3) and three items representing key AML issues (labelled Q4 to Q6).

For instance, the idea that AML has changed the job role of front-line staff (Q4) is associated with a particular question around CRM Implementation: The Company – the emphasis is on customer-driven/life event-led marketing rather than transaction-driven marketing (Q1; r = 0.284, p = 0.024).

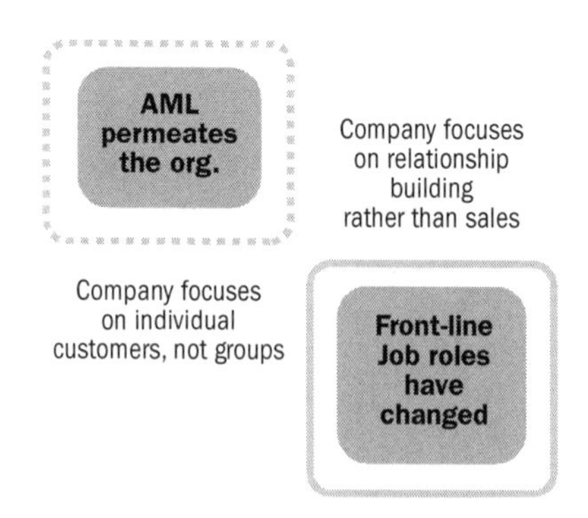

Figure 6.4. Strategic customer focus correlations.

The notion that the organisation's approach to dealing with customers has changed as a result of AML (Q5) is associated with a particular question around CRM Implementation: The Company – the company always focuses on the individual rather than customer groups (Q2; $r = 0.284$, $p = 0.024$). Moreover, the idea that AML permeates all parts of the organisation (Q6) is associated with a particular question around CRM Implementation: The Company – the company focuses on relationship building rather than increasing sales volumes as the route to competitive advantage (Q3; $r = 0.33$, $p = 0.007$).

Exploring the relationship between *AML competence* and the management and behaviours of *staff* in CRM initiatives.

Statistically significant correlations (as reported in Table B5, Appendix B) were found between two items representing key staff issues in the implementation of CRM (labelled Q1 and Q2) and three items representing key AML issues (labelled Q3 to Q5).

First we note that two key aspects of change relating to AML are supported by a single item around the nature of objective-setting for staff. The notion that staff training has changed as a result of AML (Q3) and the notion that the frequency of customer contact has changed as a result of AML (Q5) are both associated with a particular question around CRM Implementation: The Staff, namely that senior management sets objectives which reflect the company stance on CRM (Q1) (for Q3, $r = 0.267$, $p = 0.027$; and for Q5, $r = 0.262$, $p = 0.035$).

The analysis also reveals that the idea that the data accessible to front-line staff has changed as a result of AML (Q4) is associated with two questions around CRM Implementation: The Staff – first, Q1 (senior management sets objectives which reflect the company stance on CRM, $r = 0.404$, $p = 0.001$); and second, Q2 (staff often use day-to-day contacts with customers as a market research opportunity, $r = 0.298$, $p = 0.018$).

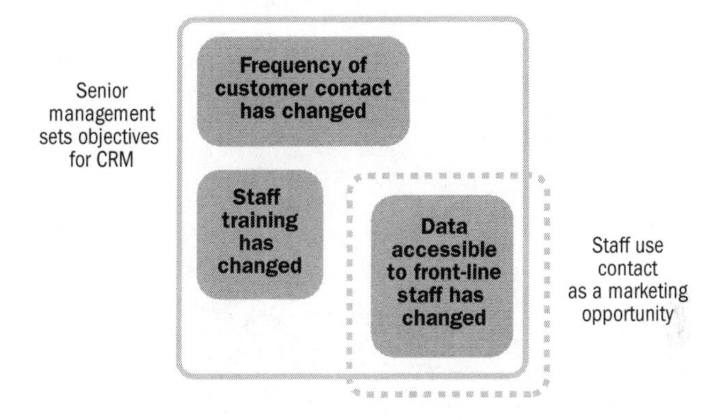

Figure 6.5. Staff correlations.